International Symposium on

Adriamycin

Edited by

S. K. Carter A. Di Marco M. Ghione
I. H. Krakoff G. Mathé

With 131 Figures

Springer-Verlag Berlin Heidelberg New York 1972

STEPHEN K. CARTER, M.D., Chief, Cancer Therapy Evaluation Branch National Cancer Institute, Bethesda/MD/USA

A. DI MARCO, M.D., National Cancer Institute, Milan/Italy

M. GHIONE, M.D., Chief, Medical Research Farmitalia, Milan/Italy

I. H. KRAKOFF, M.D., Sloan Kettering Institute for Cancer Research, New York/USA

G. MATHÉ, M.D., Institut de Cancérologie et d'Immunogénétique, Villejuif/France

71° *Convegno Farmitalia*
International Symposium on Adriamycin
Milan, 9th — 10th September 1971

ISBN-13: 978-3-642-95229-6 e-ISBN-13: 978-3-642-95227-2
DOI: 10.1007/978-3-642-95227-2

Softcover reprint of the hardcover 1st edition 1972
 Brühlsche Universitätsdruckerei, Gießen

Foreword

The impressive advances in all branches of medical science during the first half of this century with the discovery of many chemotherapeutic or immunogenic agents gave rise to brilliant achievements in the struggle against some infectious diseases and aroused in many scientists the wishful thought that drugs for cancer therapy would, soon lead to additional great success. Notwithstanding ever-increasing worldwide endeavors, the major problems in prevention or treatment of neoplastic diseases are still unsolved. The approach to the resolution of these problems follows many different pathways.

Basic research tries to cast light on the genetic and biochemical processes underlying cell division and differentiation as well as the interactions occurring between the cell and the oncogenic stimulus, or between the neoplastic cells and the different body systems endowed with immunological reactivity.

Another line of approach, coherent with the classic basis of chemotherapy, relies upon the search for new compounds selectively blocking the multiplication of the neoplastic cells.

The remarkable progress made in treating human cancer, as a result of these efforts, has been until now ascribable chiefly to the accomplishment of the chemotherapeutic approach.

Studies on the cytostatic activity of the anthracycline antibiotics carried out over many years eventually led the investigators of Farmitalia (Milan, Italy) to discover and characterize some new compounds endowed with interesting chemotherapeutic properties against malignant neoplastic diseases.

The first compound, Daunomycin (produced by S. peucetius), was isolated contemporaneously and independently (from S. caérulerubidus) also by the Rhône-Poulenc research group (Paris, France) under the name of Rubidomycin.

By a judgment of Solomon, this antibiotic was christened Daunorubicin, but the three synonyms still survive together.

An overall survey of Daunorubicin was published by Springer Verlag in 1969.

A new compound, Adriamycin (14-hydroxy-daunomycin), recently isolated by the Farmitalia group from the metabolites of a mutant strain (S. peucetius var. caesius), was shown from the very beginning to possess interesting properties and was thoroughly studied both *in vitro* and *in vivo*. Researches on Adriamycin have been carried out in Europe, America and Asia.

The experimental and clinical results gathered from these world-wide trials have been reported and discussed during the International Symposium on Adriamycin, sponsored by Farmitalia, held in Milan on September 9—10, 1971, and the papers presented to the Symposium have been collected in this book.

As can be observed by perusing the table of contents, the study of Adriamycin has brought together endeavors on its kinetics, metabolism, and mode of action, also schedule of treatment, identification of target organs, prevention of side effects, etc., which are not limited to the horizon of a particular drug but invest the whole field of the cancer chemotherapy.

Any attempt to analyse or discuss the enormous amount of scientific data accumulated so far in the literature about the cancer problem or the characteristics of malignant cells was obviously beyond the scope of this meeting.

The participants in the Symposium pragmatically put themselves in the operational position of people devoted to improving the curve of cancer with the tools at hand today, but there is no doubt that the discussion of the work carried out with Adriamycin will eventually also stimulate a re-appraisal of the basic approach to some problems of cancer.

If similar reactions can be aroused by this book also in the reader and eventually lead to an improvement in patient treatment or in basic knowledge, this will be considered by all the workers participating in the Adriamycin project as the highest reward to their effort.

On behalf of Farmitalia Company, I have to express our gratitude to all the participants at the Informal Meeting and at the International Symposium and, on behalf of the Editorial Board, I thank Springer-Verlag for the speedy publication of the Symposium.

Oktober, 1972 MARIO GHIONE

Contents

Part V. Clinical Activity and Side Effects

Part VI. Short Communications

Short Summary of the Informal Meeting Preceeding the Adriamycin Symposium

Despite the large number of growth-inhibiting compounds isolated and proved to be active against neoplastic cells in *in vitro* cultures or in animal tumors, the final goal of a highly efficient and safe therapy of human tumor is not yet achieved.

The existence of such a discrepancy between the richness of, at least potentially, active compounds at our disposal and the relatively restricted limits within which an effective therapy can be carried out is to be ascribed to different causes.

Our basic knowledge is indeed still inadequate and, to date, the outline of a proper schedule of therapy in human tumor can be rationally approached only through a multidisciplinary comparative analysis of the data obtained from experimental and clinical trials.

An unique occasion to perform such an endeavor has been provided by a meeting of researchers coming from different European, American and Asian countries to attend the International Symposium on Adriamycin.

Adriamycin (14-hydroxy-daunomycin) met the requirements for such an attempt thanks to the extensive studies on its mechanism of action, distribution, metabolism and toxicology carried out in different experimental models, in different animal species, as well as in humans.

Some specialized topics, namely the action of adriamycin in relation to cell kinetics, the metabolism of the antibiotic with particular attention to the mechanism of the toxicity and the resistance phenomena were not included into the original program of the Symposium owing to the fact that the intention of the Symposium was to emphasize the massive amount of clinical data.

In order to discuss these interesting preclinical themes, an informal meeting was held immediately before the official opening of the International Symposium and was attended by some 30 researchers, whose names are given at the end of this summary. Their presence generated such a flow of information and specific experience that it will be possible only briefly to summarize them here. The editorial board has decided to insert some abridged contributions, such as those of BOLL, DANØ, HOSHINO, MATHÉ, WHEATLEY and YESAIR among the papers presented at the Symposium, having regretfully rejected the original idea of publishing the entire discussion verbatim.

A discussion of the data on the activity displayed by adriamycin on leukocyte cultures *in vitro* opened the meeting. It soon became apparent to everyone that great difficulties exist in establishing a correlation between the data gathered from *in vitro* research carried out on leukocyte cultures growing in presence of different concentration of adriamycin and the effect observed in blood or bone marrow leukocytes *in vivo*, after administration of the drug. GOLDIN and WHEATLEY stressed the fact that the culture is a fairly complex system and the body exhibits not only a higher but also a somewhat qualitatively different set of complexity.

The *in vitro* approach to the study of drug action is often the initial step carried out in order to direct an experimental *in vivo* program. In order to obtain reliable information for the prediction of the cell behavior *in vivo*, the experimental design of *in vitro* tests has to reproduce as closely as possible the continuous changes of the active principle concentration, occurring within the body compartment under study, according to the kinetic and metabolic characteristics of the different drugs. This effort is worth while, because the knowledge of the drug activities at the cellular levels is something that chemotherapists have to acquire in order to be able to determine what the drugs are actually doing and how to improve their usage.

The research of WHEATLEY carried out on HeLa cell cultures by using a hydroxyurea-S-phase exclusion system has clearly shown that adriamycin can act on the G_2-phase of the cell cycle. The basic lesion induced by the antibiotic in the G_2-phase resulting in a block of cell progression to the mitotic phase is still unknown. Protein and RNA synthesis are largely unaffected but, as WHEATLEY stresses, a wealth of evidence has accumulated, which suggests that during the G_2-phase, RNA synthesis is largely irrelevant in determining the progression of the cell into mitosis. On HeLa cell culture the action of daunomycin is superimposable to adriamycin, but once again these *in vitro* results do not give an immediate insight into *in vivo* condition where the activity is probably related to the metabolic activation or inactivation of the drug, and to the uptake of the drug and its metabolites by the cell.

MATHÉ asked if it is possible to affirm that daunomycin and adriamycin extend their activity also to the G_0 cells, as seems to be supported by clinical evidence. In fact, treatment with these antibiotics can induce bone marrow aplasia by destroying the G_0 bone marrow cells (leaving aside the question of whether the G_0 cell really exists or if it is in reality a cell remaining in a prolonged G_1 period).

DI MARCO recalled that these antibiotics exert a clear activity upon other kinds of cells that can be operationally considered to be in G_0-phase, namely liver and kidney cells. On the other hand, LENAZ pointed out that the cells of the upper region of the intestinal cryptae, which can be considered as G_0 cells, are less sensitive to adriamycin than the other cells of the intestinal mucosa. It is indeed possible to demonstrate that these upper region cells are able to recolonize the proliferating compartment depleted by the death of the "in cycle cells" caused by treatment with high doses of the antibiotic.

The experimental model employed by WHEATLEY appears to be a very useful one, but as DI MARCO pointed out when summarizing the work of his group, one cannot forget that, by working with thimidine-synchronized cultures, it is possible to observe the activity of daunomycin and adriamycin not only on G_2 and S phase, but also on other phases of the cell cycle.

Not only different phases of the cell cycle, but also different kinds of cells exhibit a different degree of sensitivity to adriamycin; therefore, particular attention was given to the increase of monocyte number observed by BOLL in the clot cultures treated with adriamycin. The interest of this observation (a very singular one because a similar picture never occurs in clot cultures treated with many other compounds including steroid hormones) was underlined by DI MARCO, who correlated it with the finding that adriamycin, in some experimental models at least, displays an immunosuppressive activity markedly inferior to that observed with daunomycin.

Daunomycin exerts a strong, non-specific toxic activity against leukocytes. It would be interesting to see whether adriamycin is endowed with some kind of selective toxicity towards the different types of cells involved in immunological reactivity.

The importance of the study of the immunological responsiveness was emphasized by MATHÉ, from both the experimental and clinical points of view, as it is clinically important to determine the degree of variation in patients' immunologic reactivity induced by treatment with anticancer drugs. To this effect a set of tests was established, including as many as possible of the facets of the responsiveness related to the cellular or to the humoral immunity.

This immunological check-up gives the profile of the reactivity of the patients whose variation is unpredictable, as many variables are involved, such as characteristics of the drug, dosage, schedule of treatment, previous treatment, age, stage of disease, etc.

A chemotherapist should always keep in mind the lessons that antibacterial chemotherapy experience has taught us. One of the latest is the lesson that the immunological capacity of the host is the only agent that can take care of the eradication of the pathogenic entities once they have been brought under control by the drug.

As all anticancer compounds so far tested display some immunosuppressive activity, the decisive help of the immunological system may turn out to be inadequate just at the moment when its cooperation is badly needed in order to dispose of the very tough kind of "parasite" that cancer cells are, even if they can be damaged by the action of the drug.

Up to the time when a specific antitumor drug devoid of immunosuppressive action will be available (and let us take for granted that this assumption should not be considered as a conceptual nonsense), the only reasonable thing to do is to try to establish a complex strategy based on the alternation of immunosuppressive anticancer drugs and immuno-enhancing procedures, according to the results of the immunological check-up of the patient. The characteristics of adriamycin indicate that the drug can play an advantageous role in such a strategy, but more extensive basic research is needed in order to improve clinical practice.

Not only the action of drugs in relation to cell kinetics and the different cellular subpopulations, with all its immunological implications, should be properly considered in planning a therapeutic strategy, but also the drug uptake, excretion and metabolism, which can differ enormously in different animal species and in different kinds of tumors.

The appraisal of just one kinetic parameter is not sufficient to give the whole picture. For instance, the B_{16} melanoma is more sensitive to daunomycin than the L1210 leukemia, but, as YESAIR stated, more daunomycin has been found to be taken up by the L1210 cells than by the B_{16} melanoma. That suggests that the drug, even when it achieves a relatively higher concentration in the cells of the former, is not getting to its site of action. Moreover, YESAIR's observation that B_{16} melanoma cells metabolize daunomycin to a greater extent than L1210 cells, hints that a metabolite, as it is produced in situ, can exert a higher activity than the parent drug. Drug metabolism can play a crucial role, and differences in the rate of metabolism can account for differences in activity existing between compounds chemically very similar, such as daunomycin and adriamycin. Daunomycin is metabolized by mice, rats, hamsters, dogs and humans, whereas adriamycin is metabolized by hamsters to only a limited extent, according to the data of both BACHUR and YESAIR.

Adriamycin is excreted at a slower rate than daunomycin and therefore remains for longer in tissues (at least in rats and other small animal species).

Daunomycin is rapidly metabolized by a reductase. Its major metabolite, daunorubicinol, is more long-acting than the parent compound, and its kinetics appear to correspond very much to that of adriamycin. Microsomal hydrolases cleave the glycoside linkage of daunomycin, adriamycin, and of daunorubicinol as well. The resulting aglycones, as was demonstrated for at least two of them (daunomycinone and adriamycinone), are devoid of antitumor activity, but are endowed with peculiar toxic properties. The target organs appear to be the heart and the intestine, where a derangement of the function of the intrinsic nervous system can be observed.

Daunomycin equivalents have been reported by MAZZANTI to accumulate inside the myoenteric Auerbach plexuses. Paralitic ileus is correlated with these findings. According to RAUSA, the cardiotoxicity can be prevented by treatment with strophantin.

These findings not only contributed to a better understanding of the kinetics, but opened the way to a rather diversified approach to looking at these problems by drawing attention to the role played by the enzymes involved into the metabolism of anticancer drugs.

Among the enzymes BACHUR focused in daunorubicin-reductase, the enzyme which converts daunomycin into daunorubicinol. The level of the D-reductase appears to be low in normal blood white cells, high in cells of responsive human myeloblastic leukemias, and intermediate in cells of leukemias non-responsive to daunomycin. This correlation of tumor responsiveness to the enzyme level is very exciting, though more data are required on this topic.

Still more exciting are the speculations on the functions accomplished by the D-reductase which, as BACHUR emphasized, is not a trivial microsomal drug-metabolizing inducible enzyme but a soluble constitutive enzyme, ubiquitously present (in red blood cells, too) reaching higher levels in immature leukocytes and in fast-growing neoplastic cells. The native substrate of this enzyme is still unknown. It is to be considered fortuitous that this enzyme can meet a completely foreign material, like daunomycin, and be able to convert it.

Worthy of notice is the fact that in Ehrlich ascites tumor sublines resistant to daunomycin (trained and selected *in vivo* by DANØ), there is total lack of D-reductase activity as reported by BACHUR. These sublines are cross-resistant to adriamycin, and therefore the possible interaction of this latter drug with the enzyme should be examined, even if adriamycin seems not to be metabolized by the D-reductase.

Strictly correlated with this point is the study of the binding of the anthracycline antibiotics (and of the tetracyclines, too) with some lipoproteic components of the cells.

Resistance to daunomycin cannot be explained solely on the basis of the variation of the D-reductase level; another interesting mechanism of resistance has been observed by YESAIR. Daunomycin is taken up and metabolized in the resistant cells to the same extent as in the sensitive ones, but in the former the drug and the metabolite will eventually be excreted fairly rapidly, whereas in the latter they are retained.

It is possible that the two above-mentioned mechanisms correspond to different steps of resistance; it has been shown that a high level of resistance towards daunomycin is accompanied by a cross-resistance to adriamycin and to vincristine with a collateral sensitivity towards cytarabine and BCNU, according to the data gathered by DANØ on EHRLICH ascites in mice, and from HOSHINO on L1210 leukemia ascites in mice and on YOSHIDA sarcoma in rats.

These studies brought up the question of the clinical resistance phenomena which is so much more difficult to evaluate. BONADONNA reported cases in which resistance to adriamycin was accompanied by resistance to daunomycin. Individual clinical cases resistant to daunomycin and responsive to adriamycin have been observed by SINKS, TAN, MASSIMO, and MATHÉ, but the necessary controlled clinical comparison with cross-over which would prove the point remains to be done.

KRAKOFF, LUCE, and FREI all stressed the point that it is difficult to document whether the cases which do not respond to a trial with a drug are, in fact, truly resistant.

In some of these cases it is often sufficient to repeat the treatment after a certain period of time in order to obtain a clinical response. The clinical implication of the cross-resistance between vincristine and daunomycin (or adriamycin) has been discussed by CARTER, who remembered that it is a common observation that a remission in leukemic patients not responsive to a combination with vincristine and prednisone can be induced with daunomycin.

On the other hand, a disappointing aspect of the therapy of childhood leukemia is the apparent absence of synergism between vincristine and daunomycin. This can perhaps be explained by the hypothesis that a mechanism of membrane transport common to both drugs exists. MATHÉ suggested that a sequential treatment can give better results than the simultaneous application of vincristine and daunomycin.

MATHÉ and HOSHINO underlined the basic differences existing between the experimental and clinical systems. This point of view was not shared by GOLDIN, who claimed that it is just a question of sophistication in terms of experimental design, in order to bring the experimental model as close as possible to the clinical system.

The animal models are irreplaceable, either for the cross-resistance study or for the determination of another point of interest, namely for determining how many resistant cells are needed in a population of tumor cells in order to get an expression of resistance. Experimental evidence shows that 1% of resistant cells is enough.

The animal models are of fundamental importance for the study of drug interaction and to establish a rational approach to combination therapy. The conclusive remarks of CARTER and MATHÉ on this theme predict that the outgrowth of this experimental evidence, blended with the knowledge acquired from the studies on cell and drug kinetics and drug metabolism by the different tumors, will be the rational basis for the design of a suitable treatment schedule that will be a non-empirically based protocol.

MATHÉ warned that, without a sound groundwork, the haphazard application of chemotherapeutic cocktails may kill more patients than it saves.

The difficulties of the problems connected with cancer therapy emerged clearly during the meeting. The solution of those problems calls for cooperation between experts in many fields. To ensure that this cooperations, as well as an international exchange of experience and information, will be as wide and up-to-date as possible, this meeting, whose principal arguments are summarized above, and the Symposium, whose papers are collected in this book, have been organized. A great contribution to the success of this workshop was made by the chairmen of the sessions: E. POLLI, S. K. CARTER, and G. MATHÉ.

Thanks to them and to all participants in these conventions, another step has been made towards our goal.

Mai, 1972 MARIO GHIONE

List of Participants

ARCAMONE, F., Farmitalia, Research Institute, Milan (Italy)

BACHUR, N. R., Laboratory of Pharmacology, Biochemistry Section, Baltimore Cancer Research Center, National Cancer Institute, National Institutes of Health Baltimore (USA)

BONADONNA, G., National Cancer Institute, Milan (Italy)

BOLL, I., I. Med. Department of the Municipal Hospital Neukölln Berlin 47 (Germany)

CARTER, S. K., National Cancer Institute, Bethesda (USA)

DANØ, K., Department Internal Medicine, The Finsen Institute, Copenhagen (Denmark)

DAVINI, A. M., Farmitalia, Medical Research, Milan (Italy)

DEL TACCA, M., Institute of Pharmacology, University of Pisa (Italy)

DI FRONZO, G., National Cancer Institute, Milan (Italy)

DI MARCO, A., National Cancer Institute, Milan (Italy)

FREI, E., Department of Developmental Therapeutics, the University of Texas — M. D. Anderson Hospital and Tumor Institute, Houston (USA)

FUJITA, H., National Cancer Hospital, Tokyo (Japan)

GHIONE, M., Farmitalia, Medical Research, Milan (Italy)

GOLDIN, A., National Cancer Institute, Bethesda (USA)

HOSHINO, A., Laboratory of Chemotherapy, Aichi Cancer Center Research Institute, Nagoya (Japan)

HARA, Y., Cancer Center Niigata Hospital, Niigata (Japan)

KIMURA, K., National Cancer Center Hospital, Tokyo (Japan)

KONDO, M., Nippon Farmitalia, Tokyo (Japan)

KRAKOFF, I. H., Division of Chemotherapy Research, Sloan-Kettering Institute for Cancer Research, New York (USA)

LENAZ, L., Farmitalia Research Institute, Milan (Italy)

LUCE, J. K., Department of Developmental Therapeutics, the University of Texas — M. D. Anderson Hospital and Tumor Institute, Houston (USA)

MASSIMO, L., Pediatric Clinic, University of Genoa (Italy)

MATHÉ, G., Institute de Cancérologie et d'Immunogénétique, Hôpital Paul-Brousse, Villejuif (France)

MAZZANTI, L., Institute of Pharmacology, University of Pisa (Italy)

OTA, K., Aichi Cancer Center, Nagoya (Japan)
PAVONE, M., Clinic of Urology, University of Palermo (Italy)
PRAGA, C., Medical Clinic, University of Milan (Italy)
RAUSA, L., Institute of Pharmacology, University of Palermo (Italy)
RIMOLDI, R., Clinic of Phthisiology, University of Milan (Italy)
SINKS, L. F., Department of Pediatrics — Roswell Park Memorial Institute Buffalo, New York (USA)
WERNEKKE, G., I. Medizinische Abteilung des Städt. Krankenhauses, München-Schwabing (Germany)
WHEATLEY, D. N., Department of Pathology, University of Aberdeen (Great Britain)
YESAIR, D. W., Arthur D. Little Inc., Cambridge, MA (USA)

Part I
Chemistry of Adriamycin

Structure and Physicochemical Properties of Adriamycin (Doxorubicin)[1]

F. ARCAMONE, G. CASSINELLI, G. FRANCESCHI, S. PENCO, C. POL, S. REDAELLI, and A. SELVA

Istituto Ricerche Farmitalia, Milan/Italy; Istituto di Chimica del Politecnico, Centro del C.N.R. per lo studio delle sostanze organiche naturali, Milan/Italy

With 13 Figures

Adriamycin (synonym doxorubicin) is a glycoside antibiotic whose structure and stereochemistry are represented in Fig. 1 (formula I). It consists of the tetracyclic quinoid aglycone adriamycinone (14-hydroxydaunomycinone) linked to the amino-sugar daunosamine. The chemical name for adriamycin is therefore the following: (7S:9S)-9-hydroxyacetyl-4-methoxy-7,8,9,10-tetrahydro-6,7,9,11-tetrahydroxy-7-0-(2',3',6'-trideoxy-3'-amino-α-L-lyxohexopyranosyl)-5,12-naphthacenedione.

Adriamycin is prepared either by aerobic fermentation of *Streptomyces peucetius* var. *caesius* followed by solvent extraction and chromatographic purification [1, 2] or by chemical synthesis from daunomycin (synonym daunorubicin) [3, 4]. The antibiotic is crystallized as the hydrochloride from methanol-propanol mixtures.

Adriamycin is one of the daunomycin (daunorubicin) related antibiotics which are produced by selected strains of *S. peucetius* and which represent a subgroup of the anthracyclines. These are metabolites of *Streptomyces sp.* characterized by a glycosidic structure in which a tetrahydronaphthacenequinone chromophore is linked to one or more sugar residues [5]. Daunomycin structure is represented in II [6], while other daunomycin related antibiotics are dihydrodaunomycin (III) [7], compound 13 213 R.P., which consists of daunomycin linked to an unidentified sugar [8], and rubomycin B, which contains a trideoxyhexose, tentatively identified as rhodinose, linked to the daunomycin molecule [9]. A typical anthracycline, rhodomycin, is also shown (IV).

Structural Studies

Mild acid hydrolysis of adriamycin (Scheme 1) gave the water insoluble aglycone, adriamycinone (V), and an aminosugar which was isolated as the hydrochloride,

1 The trademark of Farmitalia for adriamycin is Adriblastina.

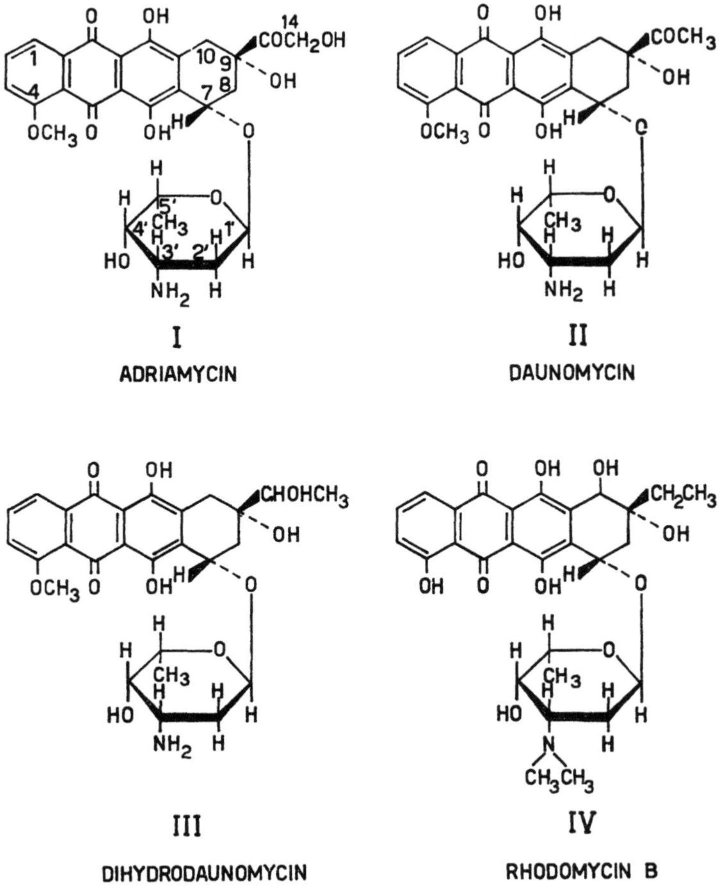

Fig. 1. Structures of *S. peucetius* derived antibiotics and of a typical anthracycline (rhodomycin B)

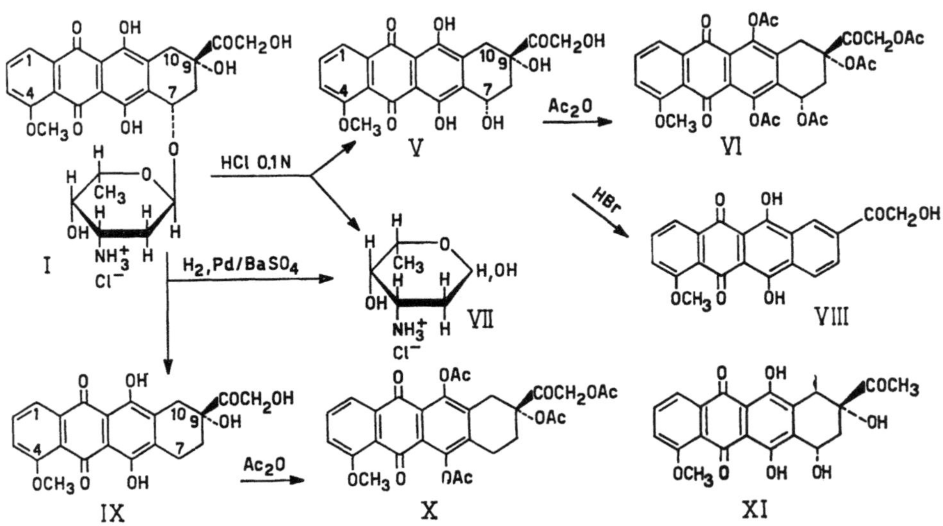

Scheme 1. Chemical degradation of adriamycin

and identified with daunosamine, or 3-amino-2,3,6-trideoxy-L-lyxohexose (VII) [10], by direct comparison (analitical data, optical rotation, chromatographic behaviour) with an authentic sample, as well as by the melting point and the infrared spectrum of the N-benzoylderivative [2].

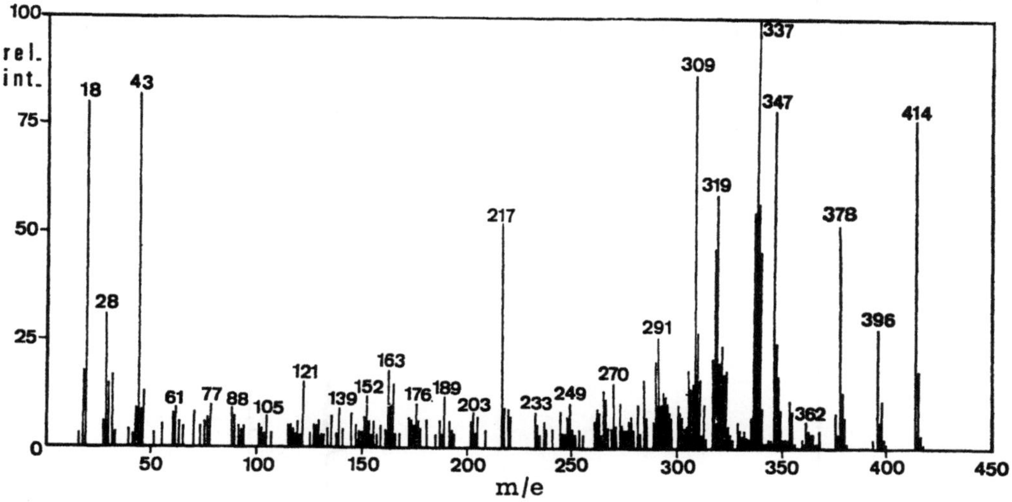

Fig. 2. Mass spectrum of adriamycinone

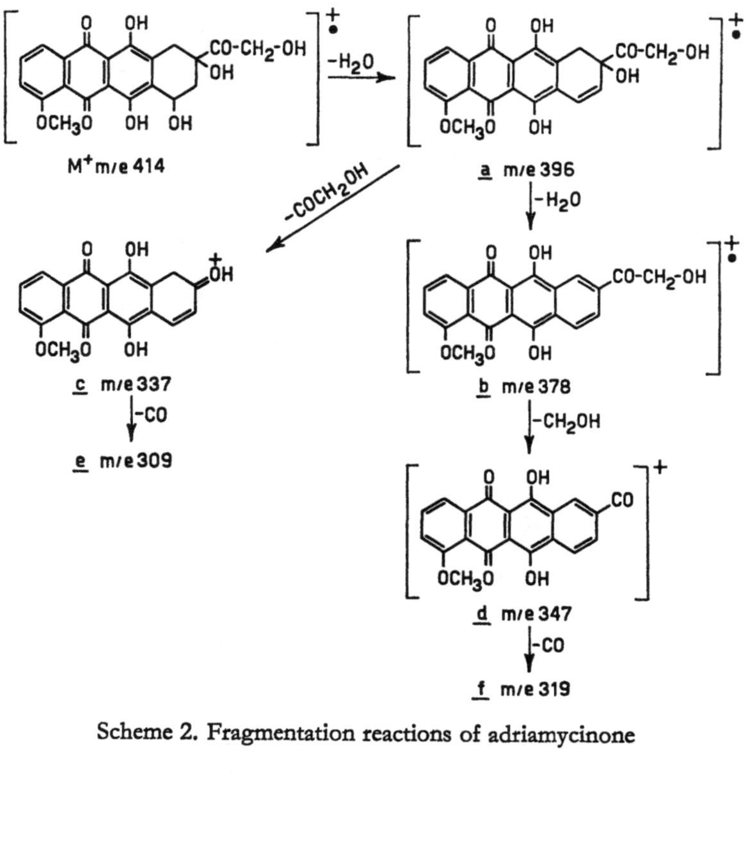

Scheme 2. Fragmentation reactions of adriamycinone

Adriamycinone (V), one OCH_3, exhibited the same ultraviolet and visible spectra as daunomycinone (XI) [6, 11] thus indicating the same type of substitution of the anthraquinone chromophore as in the latter compound. The non conjugated carbonyl band in the infrared (1727 cm^{-1}) was shifted of 9 cm^{-1} when compared with the same absorption in daunomycinone (1718 cm^{-1}). Treatment of adriamycinone with hydrogen bromide in acetic acid gave a dehydration product (VIII) whose visible spectrum was identical to that of bisanhydrodaunomycinone [6]. However, the formation of pentaacetyladriamycinone (VI), upon acetylation of V with acetic anhydride and pyridine, indicated the presence of an additional hydroxyl in respect

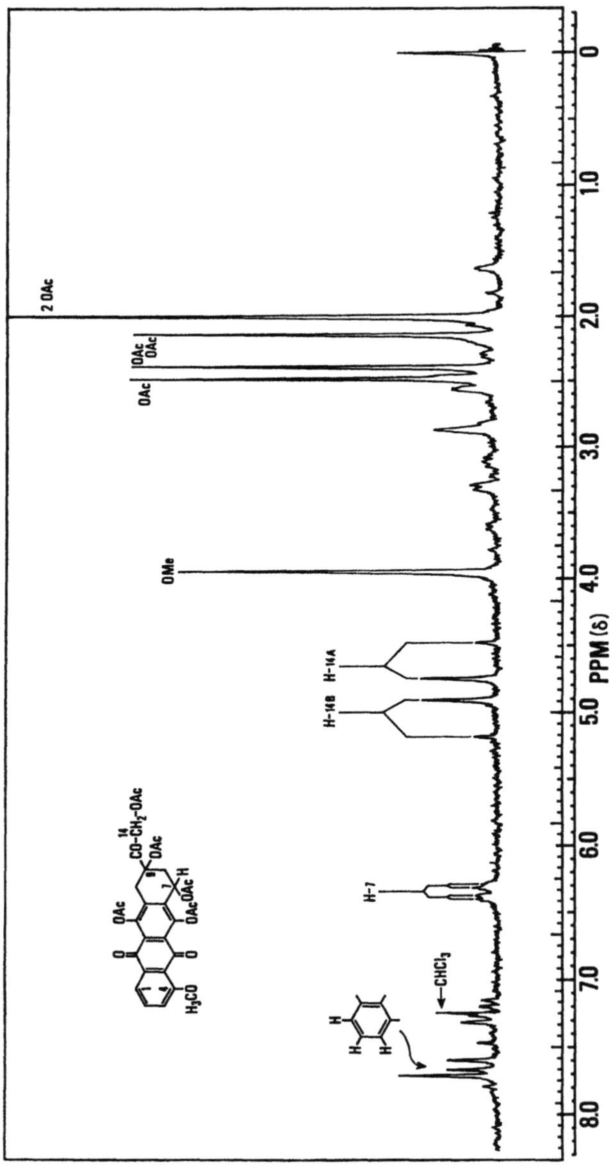

Fig. 3. N.M.R. spectrum of adriamycinone pentaacetate

to daunomycinone. The structure of V was deduced from the comparison of its mass spectrum with that of daunomycinone and from the n.m.r. spectrum of VI [13].

The mass spectrum of V is reported in Fig. 2. The molecular ion of V, m/e 414, indicated as M+ (Scheme 2), eliminates successively two moles of water giving two fragments *a*, m/e 396, and *b*, m/e 378. Fragment *a* loses the $COCH_2OH$ grouping by a simple and very favourite cleavage of the allylic bond between the ketone carbonyl and the quaternary carbon atom, yielding the ion *c*, m/e 337. Fragment *b* produces *d* by loss of 31 mass units (CH_2OH), in full agreement with proposed structure V.

The 60 Mc nuclear magnetic resonance spectrum of VI is reported in Fig. 3. The spectrum shows sharp singlets at δ 2.01 (6 H, two aliphatic $OCOCH_3$), δ 2.13 (3 H, aliphatic $OCOCH_3$), δ 2.40 (3 H, aromatic $OCOCH_3$), δ 2.48 (3 H, aromatic $OCOCH_3$), δ 3.95 (3 H, aromatic OCH_3). Four protons in the range 1.8–3.7 δ are attributed to CH_2-8 and CH_2-10. A pair of doublets (δ 4.66 and δ 5.00, 2 H, $J_{gem.}$ = 16.5 c.p.s.) is consistent with the presence of the $COCH_2OAc$ group. A double doublet at δ 6.35 (1 H, $J_{H\text{-}7,H\text{-}8_A} \sim 5$, $J_{H\text{-}7,H\text{-}8_B} \sim 1.5$ c.p.s.) indicates the C-7 benzylic $CHOCOCH_3$. The three aromatic protons appear as a complex multiplet at δ 7–8.

The position of the glycosidic linkage in I was proved by catalytic hydrogenation of I to give daunosamine and 7-deoxyadriamycinone, IX. The structure of IX was deduced, *inter alia*, by the analysis of its mass spectrum (Fig. 4, Scheme 3) and by the n.m.r. spectrum of the tetracetyl derivative X [13].

The absolute stereochemistry at C-7 and C-9 was established to be identical to that of daunomycin i.e. 7(S), 9(S), [12] on the basis of circular dichroism studies, and that at C-1′ as 1′(R) (α-glycoside) on the basis of the optical rotation data as well as of biogenetic arguments [13].

The stereochemistry was proved unequivocally by the chemical synthesis of adriamycin from daunomycin. A procedure for the chemical conversion of daunomycin to adriamycin is illustrated in Scheme 4, in which the anthraquinone part of the molecule has been omitted. Daunomycin was converted to its N-trifluoroacetyl derivative (XII), which was treated with iodine and calcium oxide in anhydrous

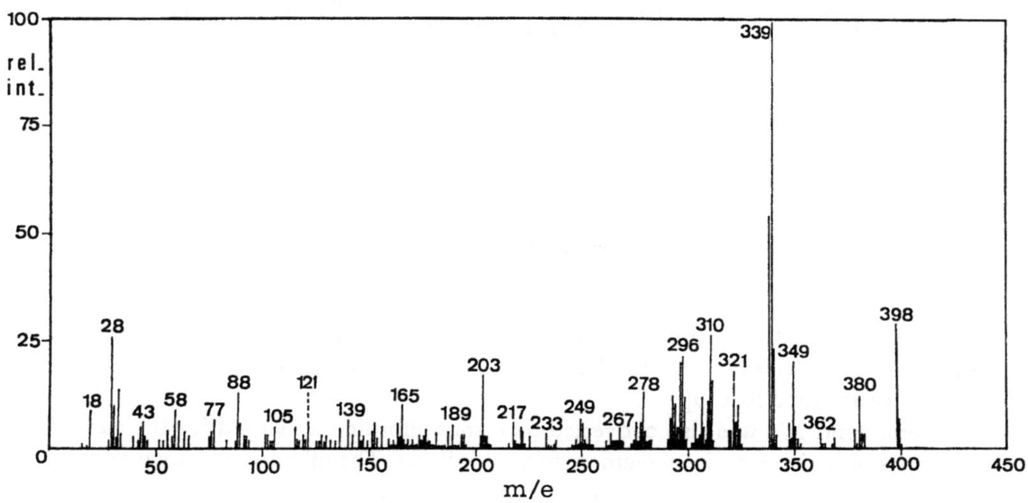

Fig. 4. Mass spectrum of 7-deoxyadriamycinone

tetrahydrofurane to give 14-iodo-N-trifluoroacetyldaunomycin (XIII). XIII was converted to 14-acetyl-N-trifluorocetyladriamycin (XIV) which, on treatment with aqueous bicarbonate, gave N-trifluoroacetyladriamycin (XV), also obtained from

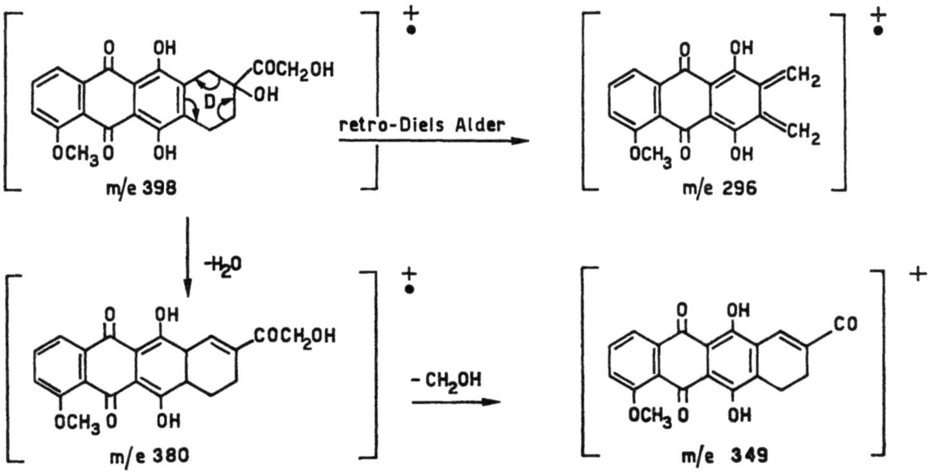

Scheme 3. Fragmentation reactions of 7-deoxyadriamycinone

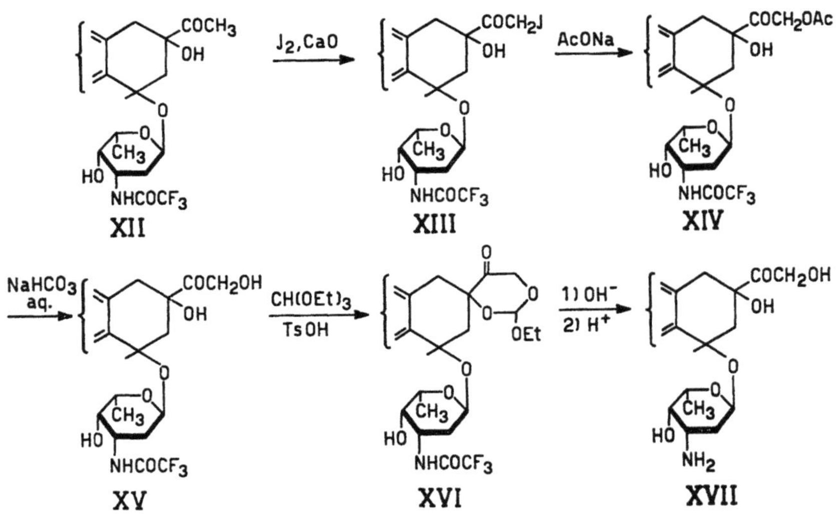

Scheme 4. Chemical correlation of adriamycin with daunomycin

adriamycin with trifluoroacetic anhydride in chloroform. This reaction sequence proves the relationship between I and II, and therefore the structure and stereochemistry assigned to I [3].

Compound XV was also converted to I, thus affording a method for the partial synthesis of adriamycin from daunomycin, as follows. The dihydroxyacetone grouping of XV was protected as the cyclic ortoformate XVI. This protection was necessary owing to the great instability of the side chain of I to the alkaline conditions used for the removal of the N-trifluoroacetyl group. Hydrolysis of XVI with 0.1 N sodium hydroxide for 20 min at 20°, followed by treatment with 1 N sulphuric acid in aqueous acetone (16 h at 20°) afforded I, *via* XVI, as the hydrochloride, identical in all respects to the natural compound [3].

Physicochemical Properties

1. Solubility

Adriamycin hydrochloride is readily soluble in water (solubility greater than 10% by weight), physiological saline and methanol, but slightly soluble or insoluble in less polar organic solvents.

2. Spectroscopic Properties

Owing to the nature of the chromophore, adriamycin shows characteristic ultraviolet, visible and fluorescence spectra. Its ultraviolet and visible spectrum in methanol (Fig. 5) shows the following maxima (extinction values are given for the

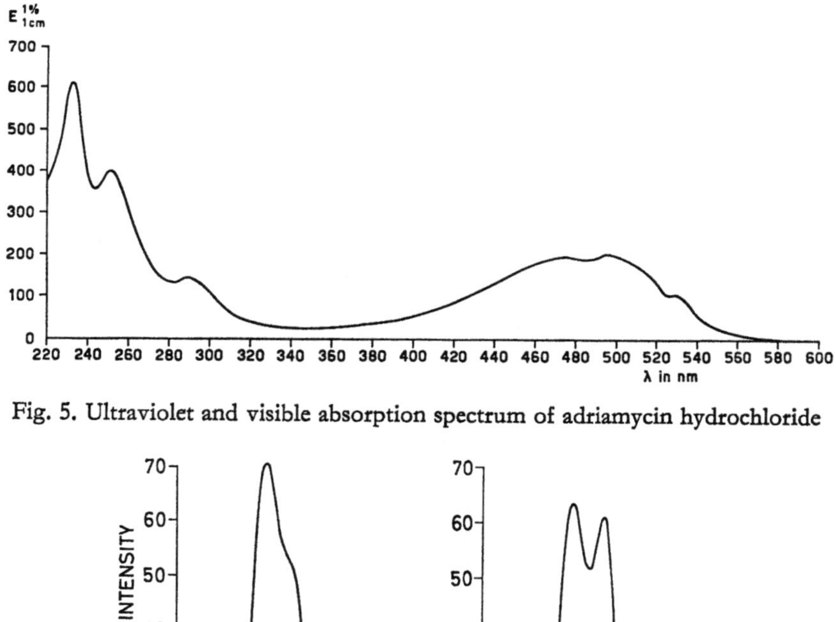

Fig. 5. Ultraviolet and visible absorption spectrum of adriamycin hydrochloride

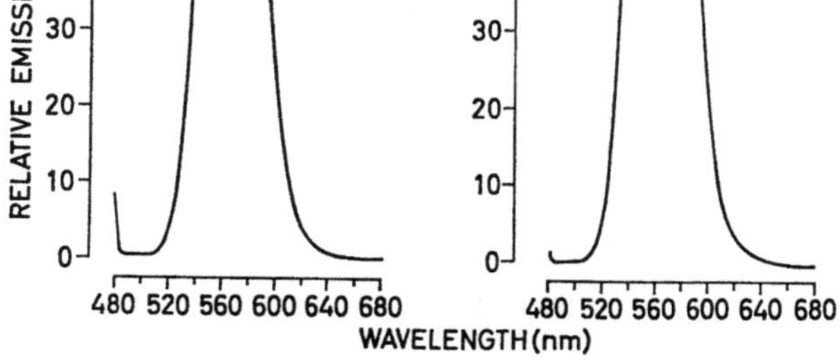

Fig. 6. Fluorescence spectra of adriamycin hydrochloride in water (left) and in ethanol (right). Concentrations were approximately 5 p.p.m.

product dried to constant weight at 56° and 0.03 Torr.); 233 nm ($E^{1\%}_{1\,cm} = 658$), 253 nm ($E^{1\%}_{1\,cm} = 440$), 290 nm ($E^{1\%}_{1\,cm} = 145$), 477 nm ($E^{1\%}_{1\,cm} = 225$), 495 nm ($E^{1\%}_{1\,cm} = 223$), 530 nm ($E^{1\%}_{1\,cm} = 124$). The maxima of absorption are shifted to

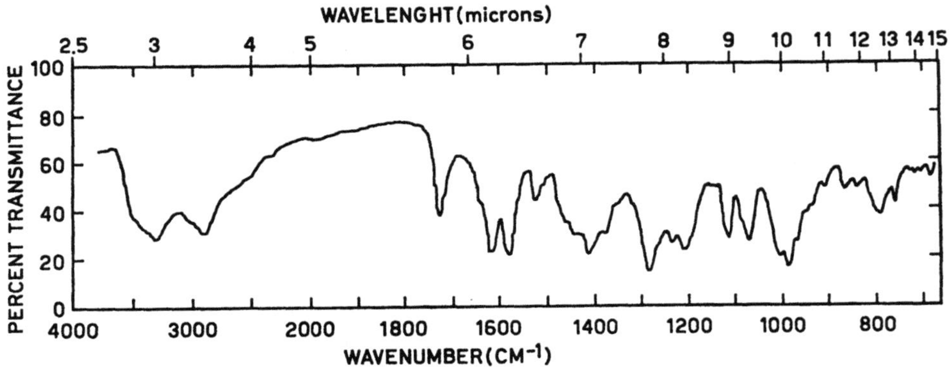

Fig. 7. Infrared spectrum of adriamycin hydrochloride (KBr pellets)

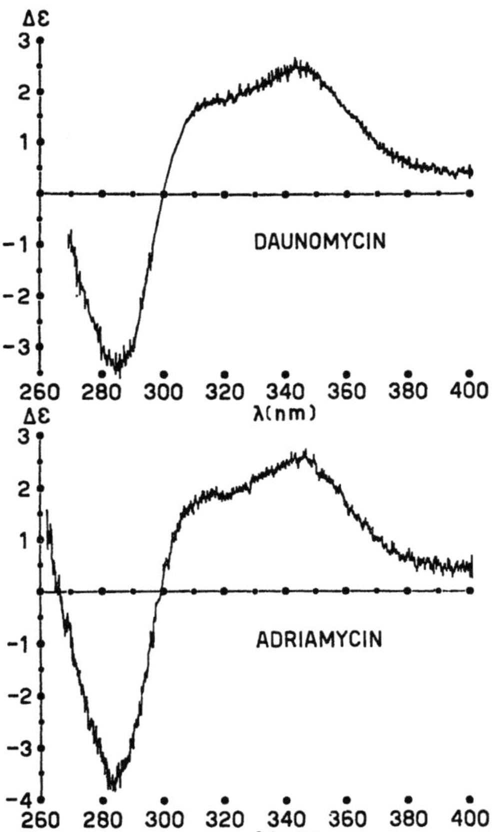

Fig. 8. Circular dichroism curves of adriamycin hydrochloride and of daunomycin hydrochloride in methanol

longer wavelengths in alkaline media. Therefore the solutions of adriamycin hydrochloride show indicator-like properties, turning from orange-red to a blue-violet colour at pH ca. 9.

The fluorescence spectra in ethanol and in water are reported in Fig. 6. The infrared spectrum is reported in Fig. 7.

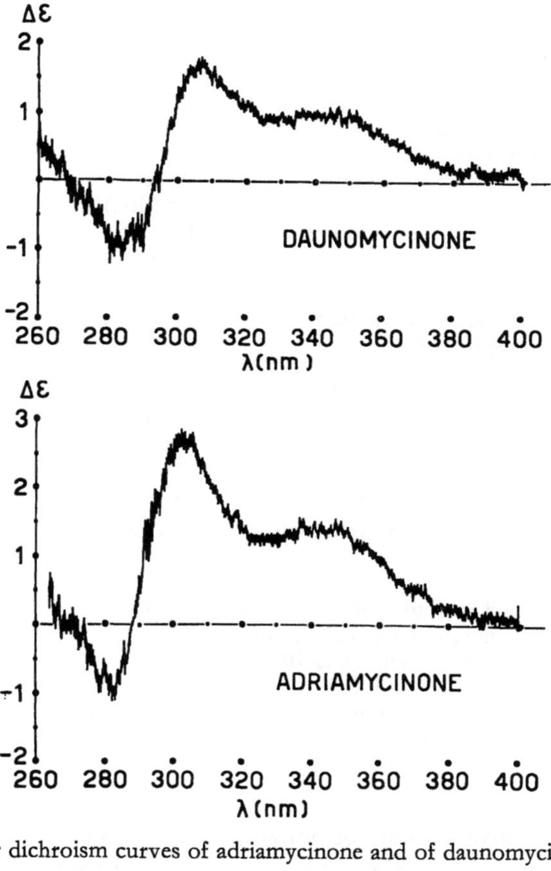

Fig. 9. Circular dichroism curves of adriamycinone and of daunomycinone (dioxane)

The chiral centres at C-7 and C-9 are responsible for the Cotton effects at 346, 315 and 285 nm shown by the circular dichroism curves of adriamycin (Fig. 8) and of adriamycinone (Fig. 9). Circular dichroism is a technique which has been proved useful for the deduction of stereochemical relationships in the field of antracyclinones [14].

3. Analytical Methods

Diagnostic data for the characterization of adriamycin hydrochloride and of adriamycinone are reported in Table 1. The antibiotic shows a high positive optical rotation which appears to the strongly dependent on the concentration. The chromatographic Rf values are in agreement with the hydrophylic character of adriamycin. In particular, adriamycin is more hydrophylic than daunomycin, as it is also shown by the direct comparison of the Rf values of the two antibiotics and of the corresponding aglycones (Fig. 10). This different behaviour is in agreement with the presence of the hydroxyl group in the side chain of I.

Table 1. Properties of adriamycin hydrochloride and of adriamycinone

	Adriamycin hydrochloride	Adriamycinone
Empirical formula	$C_{27}H_{29}O_{11}N.HCl$	$C_{21}H_{18}O_9$
Molecular weight	579.98	414.35
Percent of free base	93.72	—
pka[a]	8.22	—
Melting point	205° (dec.)	223—224°
pH of 0.5% aqueous solution	5.0 ± 0.5	—
$[\alpha]_D^{20°}$ (c 0.1)	+248°[b]	+188°[c]
Chromatographic Rf values[d]		
on paper, system A	0.1	0.3
on paper, system B	0.25	0.65
on silicagel plates, system C	0.17	0.9
D	0.33	0.8
E	0.0	0.1
F	0.0	0.25

[a] Determined on titration of adriamycin hydrochloride with N/20 sodium hydroxide.
[b] In methanol.
[c] In dioxane.
[d] A: n-butanol saturated with pH 5.4 M/15 phosphate buffer;
B: propanol, ethylacetate, water 7:1:2 by vol.;
C: methylene chloride, methanol, water 100:20:2;
D: n-butanol, acetic acid, water 4:1:5;
E: benzene, ethylacetate, petrol ether (b.p. 80—120°) 80:50:20;
F: benzene, ethylformate 1:2.

For the quantitative analysis of adriamycin, in addition to a microbiological method with *Klebsiella pneumoniae* as test organism, the spectrophotometric method is used. The latter is based on the determination of absorbance at 495 nm in methanolic solution. For the evaluation of small amounts of the antibiotic in biological fluids or in extracts the fluorimetric method has proved successful [15].

For adriamycin estimation in mixtures with other pigments, separation from these other constituents is afforded by the use of paper or thin layer chromatography.

Following the first technique on amount of solution containing from 200—500 μg of total pigments is applied on the starting line of a sheet of buffered 3MM Whatman paper and band chromatographed with solvent system propanol, ethylacetate, water 7:1:2 (by vol.) The eluates obtained by extracting the red zones appearing on the dried chromatograms with 90% aqueous methanol were used directy for the spectrophotometric reading at 495 nm.

Following the second technique thin layers of 1.0 mm thickness are prepared using Merck Kieselgel HF buffered at pH 7 with M/15 phosphate. An amount of

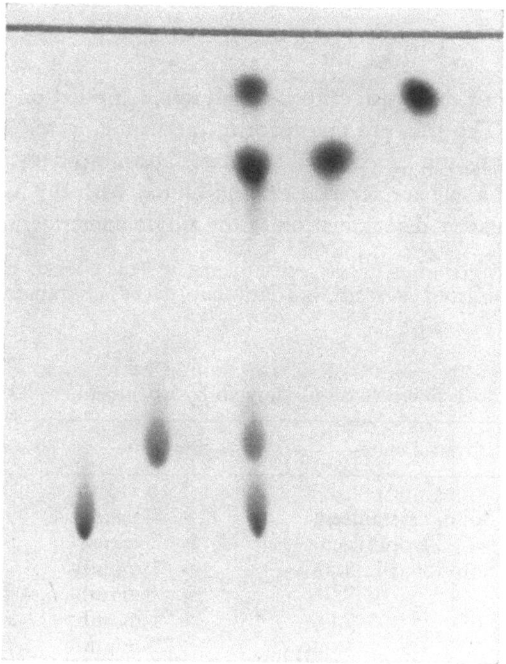

Fig. 10. Chromatographic plate showing, from left to right (single spots) adriamycin hydrochloride, daunomycin hydrochloride, adriamycinone, daunomycinone. In the middle the mixture of the four substances Silicagel plate, system C (see Table 1)

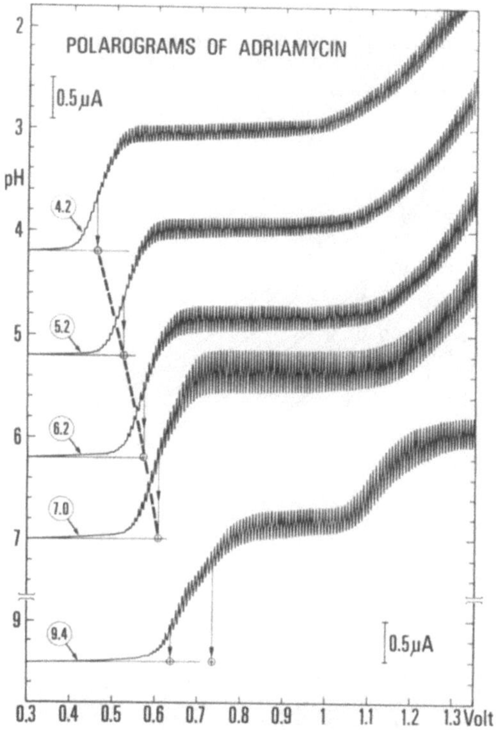

Fig. 11. Polarographic reduction of adriamycin at different pH values (Mc Ilwain buffer)

approximately 2 mg of the product to be analyzed is applied on the starting line for a length of 18 cm. The mixture chloroform, methanol, water 150:42:6 by vol. is used as developing solvent. Once dried the plates are submitted to a second development. The coloured areas are scraped off and eluted with 0.1 N hydrogen chloride in methanol. Quantitative determination of the single constituents is then carried out by spectrophotometry at 495 nm.

Owing to its quinoid system, adriamycin gives characteristic polarographic waves, as shown in Fig. 11.

Table 2. Stability of adriamycin hydrochloride at 22° C

Physical state		Stability
Solid, crystallized		> 5 years
lyophilized		> 2 years
Solution, pH	5.5[a]	> 1 month
	7.0[b]	> 1 month
	7.0[c]	> 1 month
	1 and 2[d]	< 20 hours
	3–6[e]	> 1 month
	7–8[f]	< 24 hours

[a] Adriamycin hydrochloride, 2 mg/ml, in destilled water.
[b] Same as a, pH adjusted to 7 with NaOH.
[c] Same as b, plus 0.1 or 0.2 M NaCl.
[d] Adriamycin hydrochloride, 2 mg/ml, in dilute aqueous HCl.
[e] Adriamycin hydrochloride, 2 mg/ml, in 0.06 M phosphate buffer.
[f] Adriamycin hydrochloride, 0.05 mg/ml, in 0.06 M phosphate buffer.

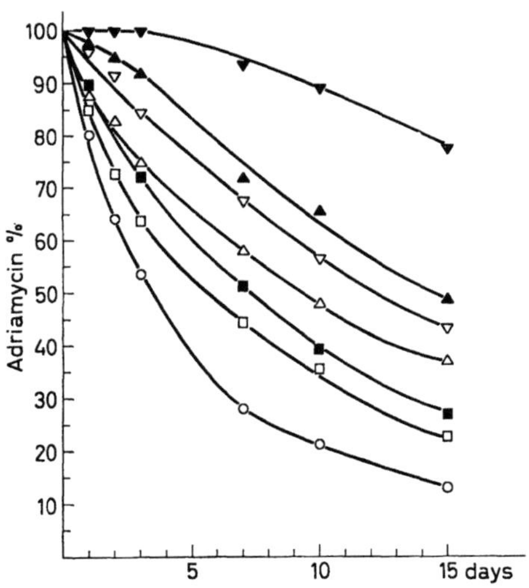

Fig. 12. Stability of adriamycin in pH 7 phosphate buffers at room temperature. Buffer concentrations: 0.06 M (○), 0.03 M (□, ■), 0.015 M (△, ▲), 0.0075 M (▽, ▼). Full symbols indicate solutions to which KCl was added in order to obtain the same ionic strength of 0.06 M buffer

4. *Stability*

Adriamycin hydrochloride is stable in the solid state, as it was stored for years at room temperature without chemical modification and without any loss of activity.

The stability of aqueous solutions of adriamycin varies with pH and with buffer concentration (Table 2 and Fig. 12) as established on the basis of spectrophotometric and chromatographic analysis.

Tritiated Adriamycin

Radioactive adriamycin was prepared from random tritiated daunomycin [16] following a procedure which involved substitution of a side chain hydrogen atom with bromine and mild alkaline hydrolysis of the brominated product [4]. Adriamycin-[^{3}H] was purified by chromatography on a cellulose column and crystallized repeatedly as the hydrochloride. The compound appeared to be radiochemically pure when analyzed by paper and thin layer chromatography (Fig. 13). The specific activity of different preparations was in the range from 7–15 mc/mM.

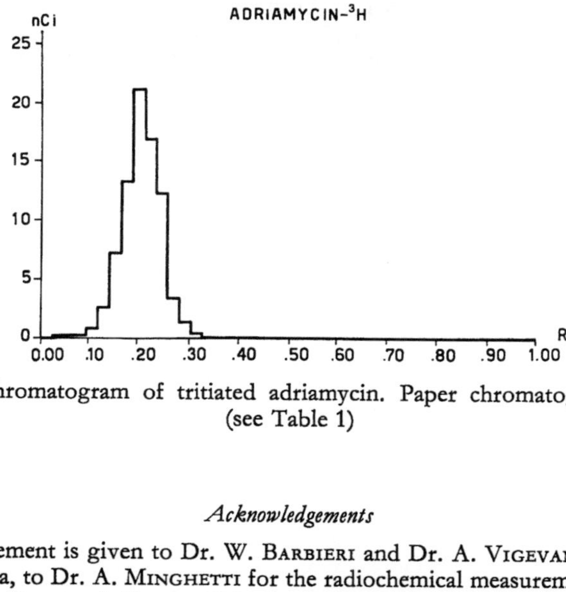

Fig. 13. Radiochromatogram of tritiated adriamycin. Paper chromatography, system B (see Table 1)

Acknowledgements

Acknowledgement is given to Dr. W. Barbieri and Dr. A. Vigevani for the infrared and n.m.r. spectra, to Dr. A. Minghetti for the radiochemical measurements, to Dr. G. B. Vicario, Istituto Donegani, Novara, Italy, for tritiated daunomycin, and to Dr. L. Valentini for the spectrofluorimetric and polarographic measurements. Expert technical assistance rendered by Sig. C. Corti, Sig. B. Pellegatta and Sig. L. Galliani is also acknowledged. The authors wish to thank professor B. Camerino for his interest and encouragement throughout this work.

References

1. Arcamone, F., Cassinelli, G., Di Marco, A., Gaetani, M.: The Antibiotic Adriamycin British Patent 1, 161, 278 (Dec. 10, 1969); Dom. Brev. Ital. 15056 A/67 (Apr. 18, 1967).
2. Arcamone, F., Cassinelli, G., Fantini, G., Grein, A., Orezzi, P., Pol, C., Spalla, C.: Biotechnol. Bioeng. **11**, 1101–1110 (1969).
3. Arcamone, F., Barbieri, W., Franceschi, G., Penco, S.: Chim. Ind. (Milan) **51**, 834 (1969).

4. Arcamone, F., Franceschi, G., Penco, S.: Procedé de preparation de l'adriamycine et de l'adriamycinone. Belg. Pat. 731, 398 (Oct. 13, 1969); Dom. Brev. Ital. 15159 A/68 (Apr. 12, 1968).
5. Brockmann, H.: Fortschr. Chem. Org. Naturstoffe **21**, 127 (1963).
6. Arcamone, F., Cassinelli, G., Franceschi, G., Mondelli, R., Orezzi, P., Penco, S.: Gazz. Chim. Ital. **100**, 949–989 (1970).
7. Arcamone, F., Cassinelli, G., Penco, S., Tognoli, L.: Substances antibiotiques et procédé pour leur préparation. Belg. Pat. 732, 968 (Nov. 13, 1969); Dom. Brev. Ital. 16449 A/68 (May 14, 1968).
8. Belloc, A., Charpentier, Y., Lunel, J., Preud'Homme, J.: Nouveau procedé de préparation de l'antibiotique 13057 R.P. French Pat. 1, 527, 892 (March 15, 1967).
9. Federova, G. B., Brazhnikova, M. Y., Mezentsev, A. S., Kshepinsky, J.: Antibiotiki **15**, 403 (1970).
10. Arcamone, F., Cassinelli, G., Orezzi, P., Franceschi, G., Mondelli, R.: J. Am. Chem. Soc. **86**, 5335 (1964).
11. Arcamone, F., Franceschi, G., Orezzi, P., Cassinelli, G., Mondelli, R.: J. Am. Chem. Soc. **86**, 5334 (1964).
12. Arcamone, F., Cassinelli, G., Franceschi, G., Orezzi, P., Mondelli, R.: Tetrahedron Letters (1968), 3353.
13. Arcamone, F., Franceschi, G., Penco, S., Selva, A.: Tetrahedron Letters (1969), 1007.
14. Brockmann, H., Brockmann, H. jr., Niemeyer, J.: Tetrahedron Letters (1968), 4719.
15. Dusonchet, L., Gebbia, N., Gerbasi, F.: Pharm. Res. Commun. **3**, 55 (1971).
16. "Rubidomycin", Bernard, J., Paul, R., Boiron, M., Jacquillat, Cl., Maral, R., ed. Recent Results in Cancer Research, Vol. 20, p. 16. Berlin-Heidelberg-New York: Springer 1969.

Part II
Activity at Cell Level

Effect of Adriamycin on the Cell Kinetic of Human Bone Marrow in vitro[1]

IRENE T. M. BOLL[2]

First Medical Department of the Municipal Hospital Neukölln, Berlin,
and Haematology Institute of the Society of Radiation Research in Association with Euratom

The effect of adriamycin was studied on *in vitro* cultured human bone marrow cells, obtained from one patient affected by chronic myeloid leukemia and from 3 patients affected by diseases which did not alter substantially the bone marrow cells. These included one myeloma patient in whom only 15% of bone marrow cells could be classified as myeloma cells, one patient suffering from liver diseases with anaemia and one patient with chronic inflammation. These cases will be regarded as non-leukemic. Bone marrow was cultivated in plasma clots, the culture medium consisting of 50% native plasm, 40% autologous serum and of 10% embryonic extract (Corhormon) [1, p. 40]. One set of cultures was untreated (controls), while one set was treated with adriamycin, at the dose of 2 µg/ml. This concentration was selected as it was claimed to induce one strong inhibition of lymphoblastic transformation and inhibition of mitosis in stimulated lymphoid cells [3], and decrease of the mitotic index as well as of the incorporation of H^3-thymidine and H^3-uridine in HeLa cells [4].

The proliferation of human bone marrow was evaluated by estimation of: mitotic index, myeloid/erythroid ratio, caryological curve [2] and maturation curve, i.e. the change of differential countings during the first three days of *in vitro* cultivation. These parameters were measured separately for erythropoietic and granulocytopoietic precursors, by counting Pappenheim-stained smears of bone marrow cells before and after 1, 2 and 3 days of cultivation. About 38000 cells were counted in the total of all the cases examined. In all the three non-leukaemic bone marrow samples investigated, the maturation of erythropoiesis was very rapid in the control cultures; therefore, the more pronounced maturation observed in the adriamycin-treated cultures could not be considered as significant. In the adriamycin-treated cultures, as compared to the controls, a decreased myeloid/erythroid ratio was observed in 2 cases after 24 h of *in vitro* cultivation, and in 1 case after 48 h, parallel

1 Supported by the Deutsche Forschungsgemeinschaft.
2 Present address: Municipal Hospital Neukölln, Rudower Straße 56, D-1000 Berlin 47.

to an enhanced number of mature erythroblasts in the differential counts at any time tested. This effect could be attributed to a disturbance by adriamycin of the enucleation of mature erythroblasts. It was reversed after 48 and 72 h of *in vitro* cultivation respectively.

More definite effects were obtained on the granulocytopoiesis, although the granulocytic maturation was not equal in the 3 experiments. The results regarding the mitotic index are shown in Table 1. The mitotic index was constantly decreased, going from $11.8^0/_{00}$ in the controls to $3.7^0/_{00}$ in the adriamycin-treated cultures after 24 h treatment, and from $7.7^0/_{00}$ in the controls to $3.0^0/_{00}$ in the treated cultures, after 48 h treatment. However, the caryological curve (percentage of mitotic phases) was not markedly changed by adriamycin treatment. Therefore, it can be concluded that, in the granulocytopoiesis of non-leukaemic bone marrow, the cell multiplication is decreased at the same extent as the mitotic index.

Table 1. Granuloblastic mitotic index ($^0/_{00}$) in human non-leukaemic bone marrow

Patient	Before cultivation	After in vitro cultivation for			
		24 h		48 h	
		Control	Adriamycin	Control	Adriamycin
RC.	15.2	9.6	0	7.0	1.9
Ch.	14.6	12.1	5.7	1.4	3.8
MÜ.	9.7	13.6	5.2	14.7	3.3
	13.2	11.8	3.7	7.7	3.0

Addition of adriamycin significantly increased binucleated granuloblasts from $1.34-2.77^0/_{00}$. These cells are probably the consequence of a defect in the cell membrane, as we previously observed after irradiation [1, p. 61].

Furthermore, the percentage of monocytes and promonocytes was hardly increased in all the investigated cases, a fact we never observed before using other (fourteen) cytostatic substances.

The experiment on the bone marrow sample obtained from a chronic myeloid leukaemia patient could be performed only up to 24 h of *in vitro* cultivation. While in the control cultures the portion of monocytes and promonocytes was 3%, in the adriamycin-treated culture this percentage increased up to 17% of all the granuloblastic cells, a value very similar to that observed in non-leukaemic bone marrow. In particular, adding adriamycin caused a decrease of the promyelocyte compartment, in favour of the myelocytes and promonocytes.

The effects of adriamycin on *in vitro* cultured human bone marrow can be summarized as follows. The regeneration of normal granulocytes *in vitro* is decreased approximately as much as that of HeLa cells [4], but not as much as that of phytohaemagglutinin-stimulated lymphocytes [3]. Binucleated granuloblasts are increased, corresponding to the giant cells observed by Massimo et al. [3]. Adriamycin is the only radiomimetic agent we investigated which enhances the monocytic precursors. The enucleation of erythroblasts seems to be disturbed. The diminution of promyelocytes observed in one case of chronic myeloid leukemia deserves further investigation in order to be confirmed.

The Author thanks B. SUCHY for helpful technical assistance.

This study was done for the Association of EURATOM and the Society of Radiation Research of Haematology (contract no. 031-64-1 BIAD).

References

1. BOLL, I.: Granulocytopoese unter physiologischen und pathologischen Bedingungen. Berlin-Heidelberg-New York: Springer 1966.
2. FIESCHI, A.: Haematologica **19**, 539 (1938).
3. MASSIMO, L., DAGNA-BRICARELLI, F., FOSSATI-GUGLIELMONI, A.: Dept. Pediat., Univ. of Genoa: Effects of adriamycin, a new antitumor antibiotic, on human lymphocytes grown in vitro. Evaluation of blastogenesis and chromosome damage at different doses.
4. SILVESTRINI, R., GAMBARUCCI, C., DASDIA, T.: Tumori **56**, 137 (1970).

Ultrastructural Alterations Induced in Hepatic Cell Nucleoli by Adriamycin[1]

GIORGIO LAMBERTENGHI-DELILIERS

Istituto di Clinica Medica I^A dell' Università di Milano
(Direttore: Prof. E. POLLI)

With 8 Figures

Introduction

During the last ten years cell lesions due to chemical compounds, and particularly to those antibiotics with antitumoral activity, have promoted numerous ultrastructural studies. As some of these agents interfere with nucleic acid synthesis, the submicroscopic alterations of the nucleus, and particularly of the nucleolus which is the source of ribosomal RNA, have been examined closely. The results lead to the conclusion that the nucleolar lesions are often typical of or even specific for a certain group of substances. Table 1 shows the three most frequent alterations seen by electron microscopy and some well-known chemical agents which determine them.

Table 1. Nucleolar alterations induced by chemical agents

Nucleolar Hypertrophy	Ethionine Isoproterenol Thioacetamide Phytohemagglutinin
Nucleolar Fragmentation	Anthramycin[a] Toyocamycin[a] Quinacrine 5-Fluorouracil Ethionine
Nucleolar Segregation	Actinomycin D[a] Anthramycin[a] Daunomycin[a] Mitomycin C[a] Mitramycin[a] Nogalomycin[a] Toyocamycin[a] Azaserine Aflatoxin 4-Nitroquinoline N-Oxide

[a] Antibiotics with antitumoral actitity.

1 This investigation was supported by CNR Grant No. 70.01530.23.17.6.3.

The aim of the present study is to report the changes in the nucleolus produced "*in vivo*" by adriamycin. The hypothesis that this organelle is morphologically compromised by the drug is based both on previous biochemical and pharmacologic studies [4–14] and on the fact that daunomycin, which is structurally related to adriamycin, produces alterations already classified as nucleolar segregation [15].

Materials and Methods

White male mice of CRF strain weighing between 20—24 g were fasted overnight prior to the intravenous injection of adriamycin. A total of 23 animals were treated: 10 received 20 mg/kg body weight of the drug dissolved in saline and 8 received 10 mg/kg body weight. An injection of saline in a volume equal to the dose of adriamycin was given to 5 control animals.

Samples for electron microscopy were removed from the left lateral lobe of the liver immediately after decapitation at the times indicated in Table 3. Small pieces of tissue were fixed in 2% cold glutaraldehyde buffered to pH 7.3 with Millonig's phosphate buffer, briefly rinsed, postfixed with phosphate-buffered 2% osmium tetroxide at 4° C, rinsed with Ringer's solution and treated with 0.5% cold uranyl acetate. Dehydration in alcohol and embedding in Epon 812 were accomplished according to the routine procedure. Thin sections were cut with diamond knives, stained with uranyl and lead and examined under a Hitachi H-11 A electron microscope. Semithin sections cut with glass knives and stained with methylene blue were studied under the light microscope in order better to localize the position in the hepatic lobule of the cell examined by electron microscopy.

Results

1. Functional Ultrastructure of Normal Nucleolus

The principal components of a normal nucleolus and their functional significance in ribosomal RNA synthesis according to the literature [2, 3, 7, 10], are presented in Table 2. The "pars chromosoma" corresponds both to the peripheral associated chromatin which tightly surrounds the nucleolus and to the intranucleolar chromatin which interdigitates with the nucleolar substance. Ultrastructural cytochemistry and

Table 2. Nucleolar components and their significance in the ribosomal RNA synthesis

Pars chromosoma Nucleolar associated and intranucleolar chromatin	=	Nucleolar DNA ↓
Pars fibrosa Fibrils 50 Å in diameter and 300–400 Å in length	=	45 S RNA ↓ 35 S 28 S and 18 S RNA ↓ Protein coat ↓
Pars granulosa Granules 150–200 Å in diameter	=	60 S and (40 S) Subribosomal particles ↓ to cytoplasm 76 S Ribosomes
Pars amorpha Amorphous proteinaceous matrix		

electron autoradiography demonstrated that this component represents nucleolar DNA producing ribosomal RNA in the form of 45 S precursor. The morphological substrate of this fraction is the "pars fibrosa" formed by fibrils of 50 Å in diameter and 300–400 Å in length. These RNA fibrils represent the origin of ribonucleoprotein granules as given in the following pathway. First the 45 S molecules are broken down into 18 S and 35 S fractions; the 18 S coated with proteins pass rapidly to cytoplasm as subribosomal 40 S particles, the 35 S become first 28 S RNA and then, after protein linkage, accumulate in the nucleolus as 60 S ribonucleoprotein particles. It is suggested that these 60 S particles constitute the "pars granulosa" formed by granules of 100–200 Å in diameter.

The fourth component is the "pars amorpha" which appears as an homogeneous matrix of protein nature. Nevertheless its functional significance is not yet clear.

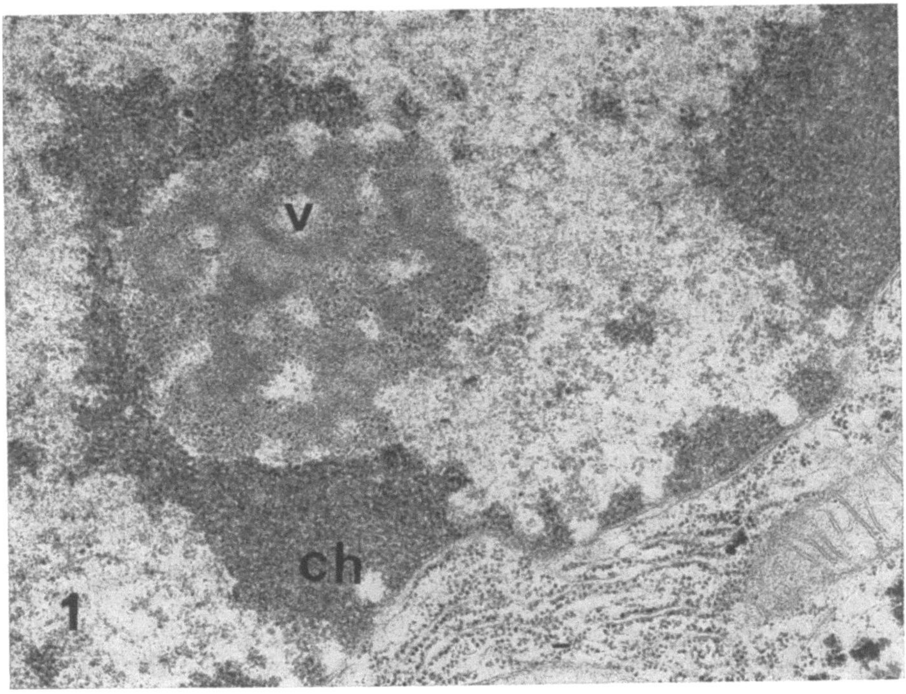

Fig. 1. Nucleolus of hepatic parenchymal cell from saline injected control mouse. Both fibrillar and granular components appear to be arranged in a trabecular structure, the nucleolonema, surrounding less dense nucleolar vacuoles (*v*). Associated chromatin (*ch*) is in close contact with the peripheral outline of the nucleolus (× 27000)

The four components seem to be constantly present in vertebrates. Their quantity and topographical distribution vary according to the cell type. In the hepatocytes of CRF mice used in this study and injected with saline, the fibrillar and granular components are arranged in a network, the nucleolonema, which surrounds electron-transparent areas also known as nucleolar vacuoles (Fig. 1).

2. Nucleolar Changes Induced by Adriamycin

Intravenous doses of 20 mg/kg body weight of adriamycin produce nucleolar fragmentation in every hepatocyte within 10 min (Fig. 2). The normal contour of the nucleolus becomes discontinuous and the trabecular structure called nucleolonema is distorted. In addition the vacuoles appear more irregular and small spherular electron-opaque masses are evident. In comparison with ethionine [13], quinacrine [5] and toyocamycin [11], adriamycin seems to produce this kind of nucleolar alteration much earlier. Nevertheless its functional significance is unknown: whether it is the result of an unspecific toxic effect of the drug or of a ribonucleoprotein synthesis inhibition it is difficult to say.

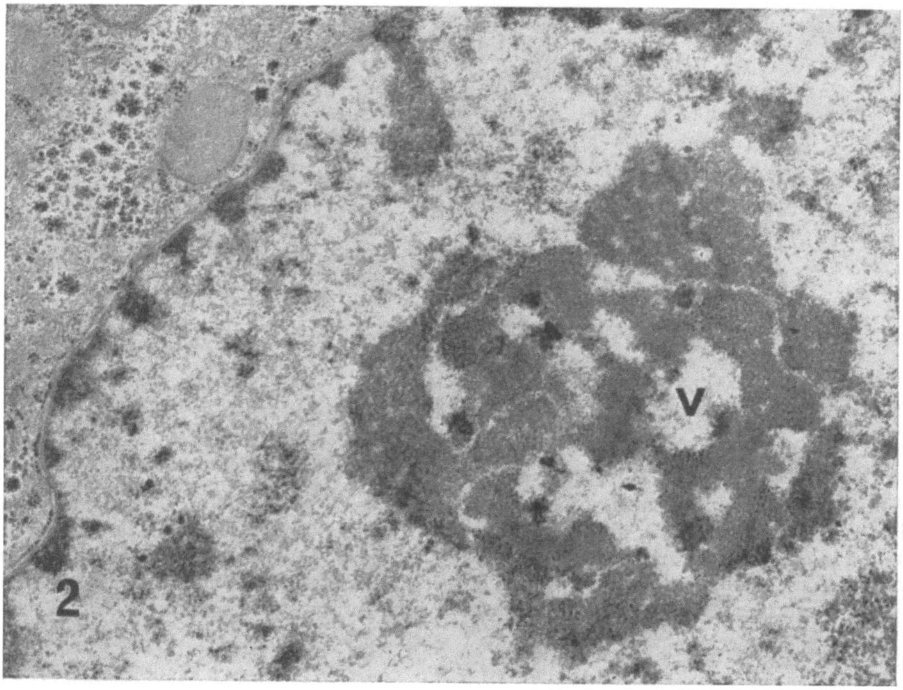

Fig. 2. Hepatic cell nucleolus of mouse 10 min after 20 mg/kg body weight adriamycin injection. The trabecular structure called nucleolonema is distorted as occurs in nucleolar fragmentation. The normal contour of the nucleolus becomes discontinuous and the vacuoles (*v*) appear more irregular (× 23000)

At 15 min, nucleolar fragmentation becomes less evident and at 30 min it is practically absent. At this time there are no differences between 20 and 10 mg/kg body weight dosages of adriamycin since both produce nucleolar segregation. This characteristic alteration was already described in detail by BERNHARD et al. [1] and by SIMARD and BERNHARD [16]: at first the nucleolus took the form of a compact sphere with the disappearence of nucleolonema and vacuoles; then came the separation and redistribution of nucleolar components into three distinct granular, fibrillar and amorphous areas; finally the progressive disappearance of the granules transformed the nucleolus into a small rounded mass mostly composed of fibrils.

Nucleolar segregation induced by adriamycin follows the same sequential steps. At 30 min after 20 or 10 mg/kg body weight injection (Fig. 3), the nucleolus has a

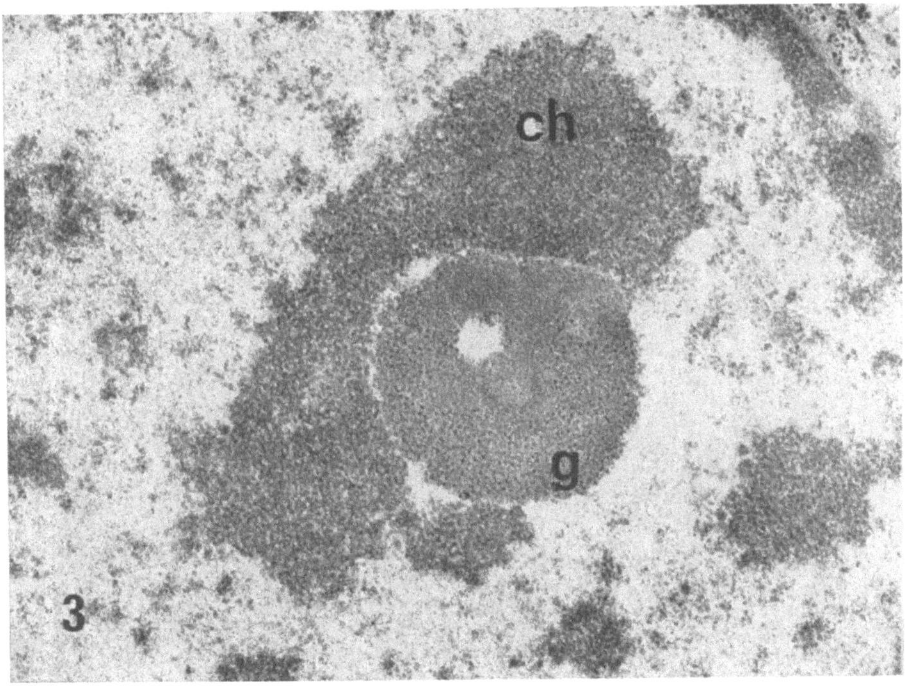

Fig. 3. Hepatic cell nucleolus of mouse 30 min after 20 mg/kg body weight adriamycin injection. The nucleolus is rounded with a more regular outline. The nucleolonema is not seen and the vacuoles are reduced. The "pars granulosa" (*g*) is separated from the fibrillar component showing the segregation phenomenon. Associated chromatin (*ch*) is separated from the nucleolus contour by a narrow electron transparent space (× 28000)

more regular contour with a greater compactness due to the loss of nucleolonema and vacuoles. The segregation phenomenon of the components is already complete but becomes increasingly evident after 1 h (Fig. 4), 2 h (Fig. 5) and 3 h (Fig. 6). According to the plane of section, different locations appear for the granular, fibrillar and amorphous components. The granules may be concentrated at opposite poles of the nucleolus or they may surround the "pars fibrosa" condensed in the central area with the "pars amorpha" occupying small peripheral areas (Fig. 5). Associated chromatin is often separated from the nucleolus by a narrow space of low electron density (Fig. 3); in addition the extension of the associated chromatin quantitatively decreases until it is scarcely visible (Fig. 4–6). At 3 h some nucleoli are surrounded by a membrane like ring of amorphous material (Fig. 6). Finally 6 h after drug injection, nucleoli are transformed into a circular mass of closely packed fibrils (Fig. 7).

What is the significance of these sequential alterations? No biochemical analyses have been done in parallel to allow correlations. Nevertheless, according to the majority of Authors [2, 3, 6, 15] nucleolar segregation is a specific lesion caused by a group of compounds which interfere with the nucleolar DNA–RNA template

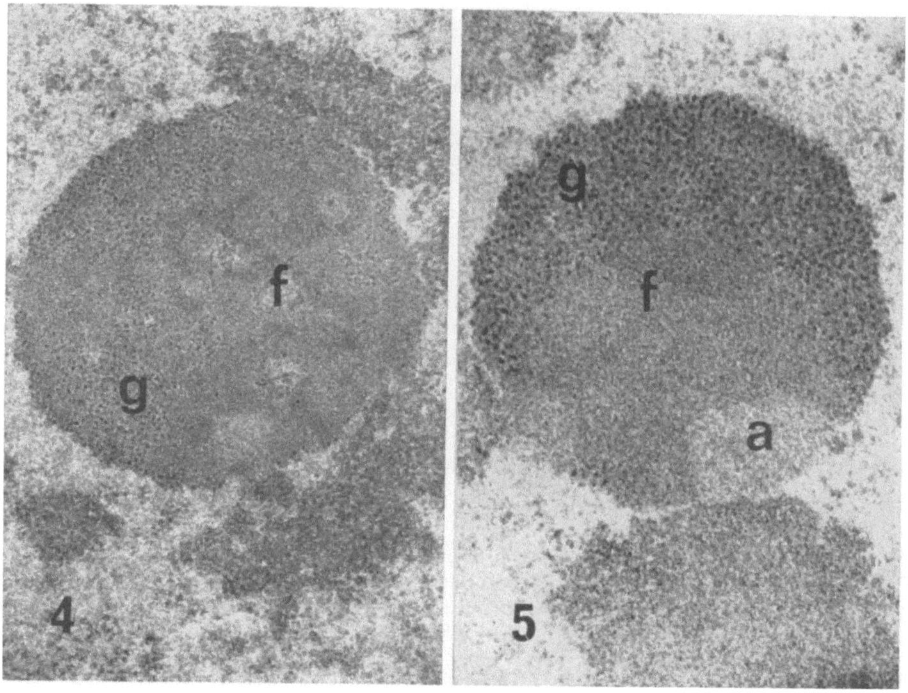

Fig. 4. Hepatic cell nucleolus of mouse 1 h after 20 mg/kg body weight adriamycin injection. The round nucleolus has a more regular contour with greater compactness due to the loss of nucleolonema and vacuoles. The fibrillar (*f*) and granular (*g*) components are clearly segregated in different areas. Associated chromatin is scarcely visible (× 30500)

Fig. 5. Hepatic cell nucleolus of mouse 2 h after 10 mg/kg body weight adriamycin injection. The nucleolus without nucleolonema and vacuoles shows a regular circular contour. Nucleolar segregation is very evident. "Pars granulosa" (*g*) surrounds "pars fibrosa" (*f*) which is centrally located. "Pars amorpha" (*a*) is concentrated in small peripheral areas. Associated chromatin is almost absent (× 42500)

system. For actinomycin D, Goldblatt et al. [6] support the hypothesis that the segregation phenomenon is a reflection of some alterations of DNA rather than the consequence of inhibition of RNA synthesis. The disappearance of vacuoles [9], the decrease of associated chromatin with the formation of an electron-transparent space [8–12] and sometimes of a pseudomembrane [17–18] around the nucleolus have also been correlated with a disturbance of nucleic acid metabolism. Therefore, from analogy, the alterations caused by adriamycin seem to be the morphological counterpart of nucleolar DNA and RNA synthesis disturbance already proved by biochemical data [4–14]. As a consequence, the maturation of fibrils into granules might be

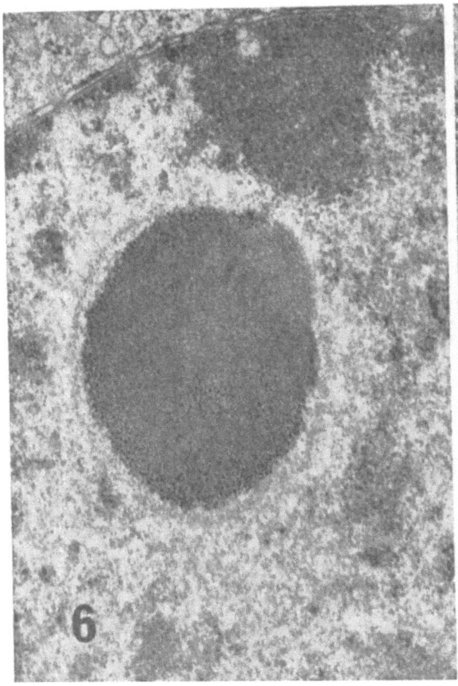

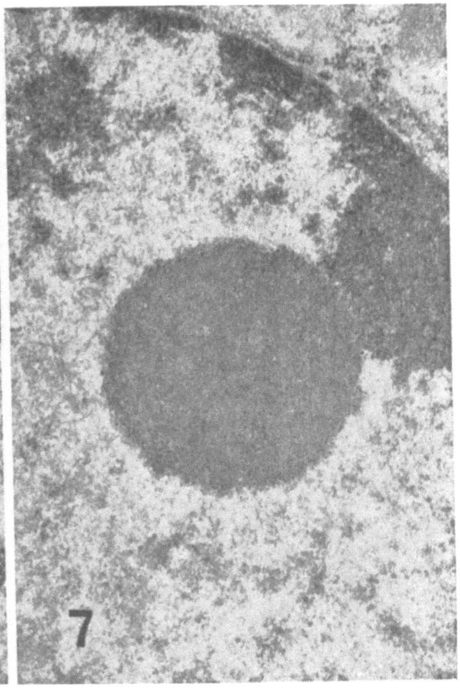

Fig. 6. Hepatic cell nucleolus of mouse 3 h after 20 mg/kg body weight adriamycin injection. The nucleolus still showing the segregation phenomenon of the components, is surrounded by a membrane-like ring of amorphous material. Associated chromatin is practically absent (× 35500)

Fig. 7. Hepatic cell nucleolus of mouse 6 h after 10 mg/kg body weight adriamycin injection. The nucleolus without associated chromatin is transformed into a circular mass of closely packed fibrils. The granular component is scarcely visible (× 30500)

Table 3. Sequential morphological alterations in hepatic cell nucleoli induced by 20 and 10 mg/kg of adriamycin

Times examined	Doses (mg/kg)	Nucleolus aspect
5 min	20	Normal nucleoli
10 min 15 min	20	Nucleolar fragmentation
30 min 1 h 2 h 3 h 6 h	20.10	Nucleolar segregation
6 h		(Some normal nucleoli)
12 h 24 h	20.10	Normal nucleoli

blocked and the nucleolus transformed into a small circular mass devoid of granular component.

As summarized in Table 3, the nucleolar lesions induced by adriamycin are rapidly reversible. In fact at 6 h after 20 or 10 mg/kg body weight drug exposure, in some hepatocytes there is an early reconstruction of the nucleolonema (Fig. 8).

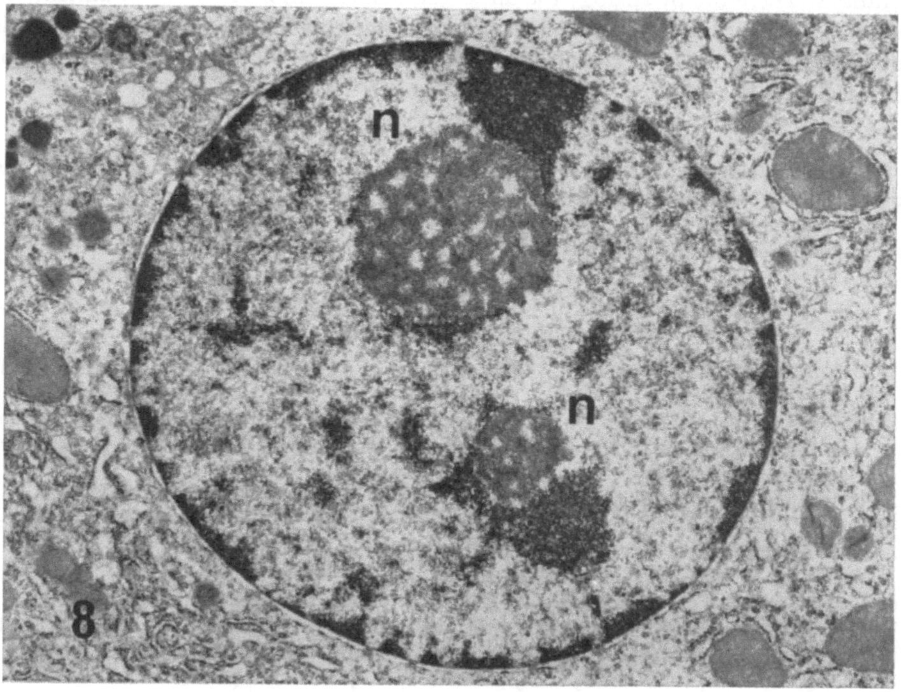

Fig. 8. Hepatic cell nucleoli of mouse 6 h after 10 mg/kg body weight adriamycin injection. The nucleoli (*n*) show a substantial return to normal structure. The nucleolonema and the vacuoles have reappeared giving the nucleolus a reticular appearance (× 14500)

At 12 and 24 h the nucleoli are indistinguishable from those of saline-injected controls. To explain this rapid recovery several mechanisms might be proposed: hepatic cell regeneration, early metabolic breakdown of adriamycin, "de novo" reorganization of nucleoli at a different locus as already suggested by GOLDBLATT et al. [6] for actinomycin D.

Summary and Concluding Remarks

1. The administration of adriamycin to CRF mice induces a sequence of ultrastructural alterations in hepatic cell nucleoli. The earliest change is the fragmentation of the nucleolonema followed by the segregation of nucleolar components.

2. The lesions present in all parts of the liver lobule are dose-dependent. Nucleolar fragmentation is in fact produced by the highest doses of adriamycin.

3. The nucleolar alterations described in this study are like those induced by other antibiotics with antitumoral activity. In particular the sequential steps of the

segregation phenomenon are very similar to those already described after actinomycin D treatment [6] both for their early appearance and for the rapid recovery of the nucleolus.

4. It is well known that the segregation phenomenon is a specific lesion produced by certain chemical compounds which, binding DNA, block the nucleolar RNA synthesis. Therefore by analogy the ultrastructural alterations induced by adriamycin might be the morphological counterpart of a nucleolar DNA and RNA synthesis inhibition. This tentative conclusion is supported by the previously reported biochemical data [4–14].

Acknowledgments

I wish to thank Prof. C. Bertazzoli of Farmitalia Research Institute, Milan, for the kind supply of adriamycin injected animals.

I am grateful also to Mrs. L. Bonacina for assistance in the preparation of this manuscript and to Mrs. Bassetti-Poirier for expert technical assistance.

References

1. Bernhard, W., Frayssinet, C., Lafarge, C., Le Breton, E.: Compt. Rend **261**, 1785–1788 (1965).
2. Bernhard, W., Granboulan, N.: Electron microscopy of the nucleolus in vertebrate cells. In: Dalton, A. J., Haguenau, F., (Eds.): The nucleus, pp. 81–149. New York: Academic Press 1968.
3. Busch, H., Smetana, K. V.: The nucleolus. New York: Academic Press 1970.
4. Casazza, A. M.: L'adriamicina: chimica, farmacologia, attività biologica. Corso Superiore Chemioterapia Tumori, Milano, 23–27 Marzo 1970, 337–355.
5. Fedorko, M. E., Hirsch, J. G.: Cancer Res. **29**, 918–924 (1969).
6. Goldblatt, P. J., Sullivan, R. J., Farber, E.: Cancer Res. **29**, 124–135 (1969).
7. Hay, E. D.: Structure and function of the nucleolus in developing cells. In: Dalton, A. J. Haguenau, F., (Eds.): The nucleus, pp. 1–79. New York: Academic Press 1968.
8. Jezequel, A. M., Bernhard, W.: J. Microscopie **3**, 279–296 (1964).
9. Johnson, J. M.: J. Cell Biol. **43**, 197–206 (1969).
10. Marinozzi, V.: J. Ultrastruct. Res. **10**, 433–456 (1964).
11. Monneron, A., Burglen, J., Bernhard, W.: J. Ultrastruct. Res. **32**, 370–389 (1970).
12. O'Hegarty, M. T., Harman, J. W.: Cancer Res. **29**, 521–528 (1969).
13. Shinozuka, H., Goldblatt, P. J., Farber, E.: J. Cell Biol. **36**, 313–328 (1968).
14. Silvestrini, R., Gambarucci, C., Dasdia, T.: Tumori **56**, 137–148 (1970).
15. Simard, R.: Int. Rev. Cytol. **28**, 169–211 (1970).
16. Simard, R., Bernhard, W.: J. Int. Cancer **1**, 463–479 (1966).
17. Stenram, U.: Z. Zellforsch. **94**, 282–292 (1969).
18. Stenram, U., Willen, R.: Cancer Res. **26**, 765–772 (1966).

Effects of Adriamycin on Blastogenesis and Chromosomes of Blood Lymphocytes: "in vitro" and "in vivo" Studies

L. MASSIMO, F. DAGNA-BRICARELLI and M. G. CHERCHI

Clinica Pediatrica "G. GASLINI" dell'Università di Genova
(Direttore: Prof. R. BULGARELLI)

With 6 Figures

Adriamycin is a new antibiotic of the anthracycline group [1, 3], analogue of daunomycin. It showed a higher therapeutic activity than daunomycin on several experimental tumors, and has been successfully tested in clinical trials on solid tumors and leukemias.

Studies on *in vitro* cell cultures demonstrated that adriamycin quickly penetrates into the cell and binds to nuclear structures, especially to the perinucleolar chromatin, causing an immediate disturbance of the mitotic process, shown by prophasic inhibition at low concentrations and, at higher doses, by mitotic block. DNA and RNA synthesis are highly impaired, as well as cell vitality [5]. VIG [6] has recently reached the same conclusions by studying the chromosomic aberrations caused by adriamycin on human lymphocytes cultures.

The antiblastic activity of adriamycin and its use in human leukemias prompted us to investigate more deeply the effects of this antibiotic on human lymphocytes. This was accomplished by studying: 1) the effect of adriamycin on normal human lymphocytes stimulated with phytohaemoagglutinin (PHA) *in vitro*; 2) the alterations caused by adriamycin treatment on lymphocytes of children suffering from solid tumors in advanced stages.

Materials and Methods

Patients

Venous blood was taken for the first experiment from 5 healthy donors, and for the second one from 8 children suffering from advanced neoplastic diseases, submitted to treatment with adriamycin. In these patients blood was taken on the day preceeding the first administration of the drug and on the day following the sixth and last dose. The drug was given at a daily dose of 0.5 mg/kg in two courses of three injections each, with a five-day rest between them. The children had never been treated before with other antiblastic drugs or radiations.

Lymphocytes Cultures

Lymphocytes were separated through sedimentation, and cultivated *in vitro*, at the concentration of 10^6 cells/ml in NCTC 109 medium, added with 20% inactivated human serum, phytohemagglutinin (PHA) and antibiotics (streptomycin 1000 μg/ml; penicillin 1000 U/ml). Lymphocyte cultures from healthy donors were added also with adriamycin at the following doses: 0 (control); 0.05; 0.1; 0.2; 0.3; 0.5; 1; 1.5; 2; 3; 5 μg/ml. Cultures were incubated at 37 °C for 72 h. At this time, after centrifugation, slides were prepared; cells were fixed and stained with May Grunwald-Giemsa. The percentage of blast cells was evaluated by counting 1000 cells for each culture.

In order to study chromosome damage, after 69 h incubation at 37 °C, a set of cultures was treated with Colcemid (40 μg/ml), then at 72 h processed as before described. The number of altered mitoses (chromosomal fragmentation, chromatin gaps, dicentric and ring chromosomes, fragments and interchanges) was evaluated on 100 mitoses per culture as well as the number of aneuploid, hypo-, hyper- and polyploid cells, as compared to normal controls.

Results

Action of Adriamycin on "in vitro" Cultured Normal Human Lymphocytes

a) Alteration of Lymphocytes Phytoresponsiveness. Adriamycin, added to normal human lymphocytes cultivated *in vitro* in the presence of PHA, caused a strong inhibition of phytoresponsiveness (Fig. 1). The inhibition was a linear function of the adriamycin dose, and was present also at the lowest dose tested (0.05 μg/ml,) At 3—5 μg/ml, no blastic transformation occurred.

At adriamycin doses at which a good responsiveness to PHA was still present, lymphocytes showed to be able to transform into lymphoblasts, but could not undergo subsequent normal divisions; this led to several abnormal pictures in the blastic cells morphology.

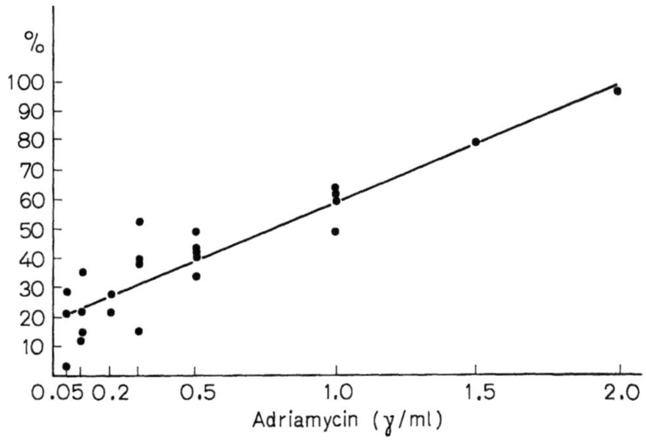

Fig. 1. Activity of adriamycin on "*in vitro*" cultured normal human lymphocytes. Abscissa: concentrations of adriamycin (μg/ml); ordinate, percentage of blastogenesis inhibition, as compared to the control

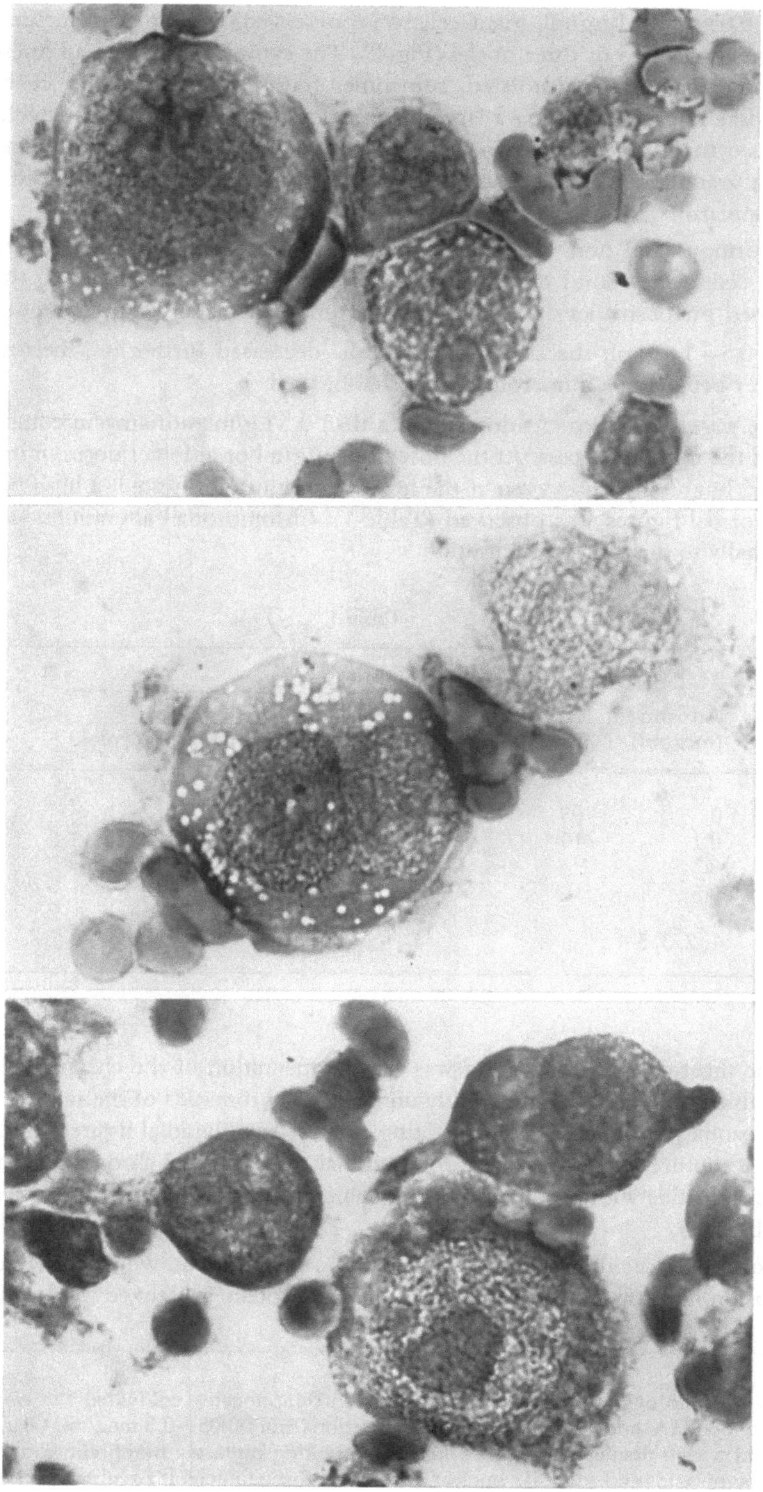

Fig. 2. (Legend see page 38)

At 0.05 and 0.1 μg/ml, giant cells were observed, having diameter up to 70 μ, and sometimes two or three nuclei (Fig. 2). The cytoplasm showed an intense basophilia and was often vacuolated, sometimes fragmented, leading to characteristic comet-like figures or leaving a large bare nucleus containing 2, 3 or 4 nucleoli. The nucleus/cytoplasm ratio was in favor of the nucleus. Nuclear chromatin showed often a granular appearance, and nuclei were found fragmented and scattered into the cytoplasm.

Enormous and non vital blastic cells were also observed. At 0.2—0.3 μg/ml, blastic cells were found having a morphological appearance similar to that above described, but of smaller size; in fact their maximum diameter was not larger than 30 μ.

At 0.5—1 μg/ml, the size of blastic cells decreased further on, their maximum diameter being 10 μ; numerous cells were not vital.

b) Cytogenetic Effects. At doses higher than 0.5 μg/ml, adriamycin completely inhibited the mitotic process. At the dose of 0.5 μg/ml or at lower doses, mitoses were present, but at any dose, even at the lowest concentration tested, a high percentage of abnormal figures was observed (Table 1). Chromosomal alterations varied proportionally to the drug concentration.

Table 1

	Mitoses %			
			Aneuploid	
Adriamycin (mcg/ml)	Normal morphology	Abnormal morphology	Hypoploid	Polyploid
0	98	—	—	2
0.1	16	74	7	3
0.3	4	78	7	11
0.5	—	90	?	10
1, 2, 3, 5	No mitoses			

The most common alteration was the fragmentation of the chromatides, whose ends joined together with apparently undamaged chromatids of the same or of other chromosomes, producing dicentrics, rings, racket or polyradial figures (Fig. 3). Centromere rupture with subsequent longitudinal separation of chromatids, rupture of both chromatids with production of fragments, rearrangements, translocations, were also observed.

Morphological alterations of chromosomes were present, which appeared to have an erosion-like aspect, due to damage of the protein sheath covering the DNA.

Fig. 2. Morphologic changes of normal human lymphocytes cultivated "*in vitro*" in the presence of PHA and adriamycin at concentrations from 0.05—0.3 mcg/ml. Giant cells are observed with a diameter up to 70 micron, presenting intensely basophilic and vacuolated cytoplasm, and granular nuclear chromatin. Large nucleoli are also observed

Adriamycin also caused stickiness and agglutination of chromosomes; polyploidy, due either to endoreduplication or to lack of cell division; aneuploidy with hypo- and hyperploidy.

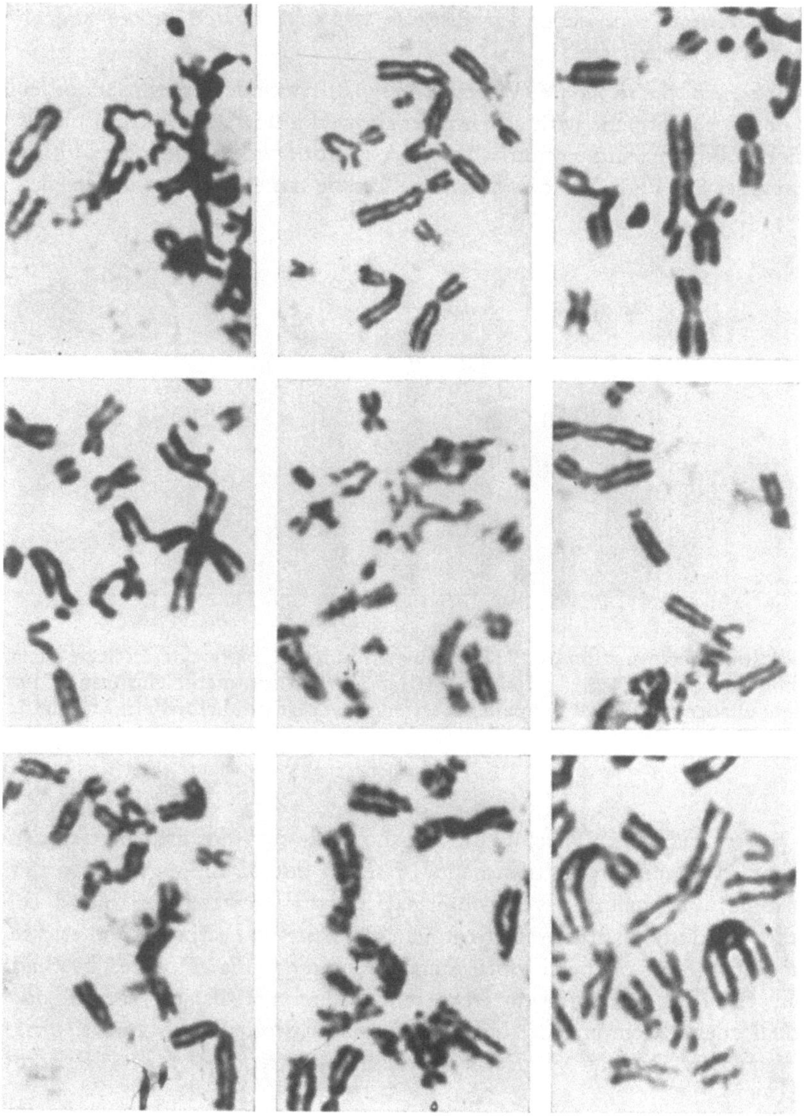

Fig. 3. Major and minor chromosome aberrations caused by adriamycin: breaks, fragments, rearrangements, chromatid exchanges and fusions

The chromosomal alterations described were more frequent in cultures treated with adriamycin at the doses of 0.3—0.5 μg/ml. At 0.1 μg/ml, adriamycin caused a high percentage of abnormal mitoses, but only 30% of these showed severe damage. Chromosomes, though altered, preserved their morphology and identity.

Action of Adriamycin Therapy on Lymphocytes of Children Suffering from Solid Tumors in Advanced Stages

The possible appearance of cell and chromosome changes in the lymphocytes of children suffering from neoplastic diseases was investigated before and after the treatment with adriamycin.

a) Alteration of Lymphocyte Phytoresponsiveness. Lymphocytes from blood collected before or after treatment with adriamycin showed a normal ability to blastic transformation following stimulation with PHA. Mitotic index remained within normal limits in all cases (Fig. 4). No spontaneous blastic transformation occurred.

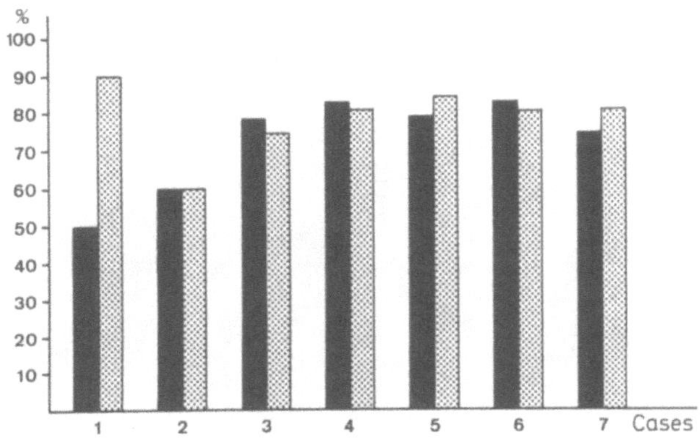

Fig. 4. Blastic transformation of PHA-stimulated human lymphocytes collected from neoplastic children before (■) and after (▒) adriamycin treatment. This test did not show any severe abnormality in all the patients tested, never treated before with antiblastic drugs or radiations

b) Cytogenetic Effects. One child did not show any chromosomal alteration, neither before nor after adriamycin treatment. In seven out of the eight cases examined chromosome abnormalities were observed in lymphocytes from blood collected both before and after adriamycin treatment. In particular, erosion and stickiness of the chromatids, lysis of the chromatin and aneuploidy (Fig. 5) were observed. Two children, the first one suffering from reticulosarcomatosis, the second one from embryonal adenocarcinoma of the testis (orchioblastoma) with spread metastases, were affected by Down's syndrome, showing a karyotype with 21-trisomy (47, XY, 21G).

In the seven cases with chromosome abnormalities a statistical analysis was performed to evaluate the significance of aneuploidy before and after adriamycin treatment.

Cells were classified in three classes: normal cells, cells with morphological changes, aneuploid cells. The data were collected in a 2×3 table, and the total χ^2 was calculated in order to investigate the correlation between the three classes of cells.

To make a comparison between the percentage of abnormal versus normal cells before and after adriamycin treatment, χ^2 was dissociated according to the method

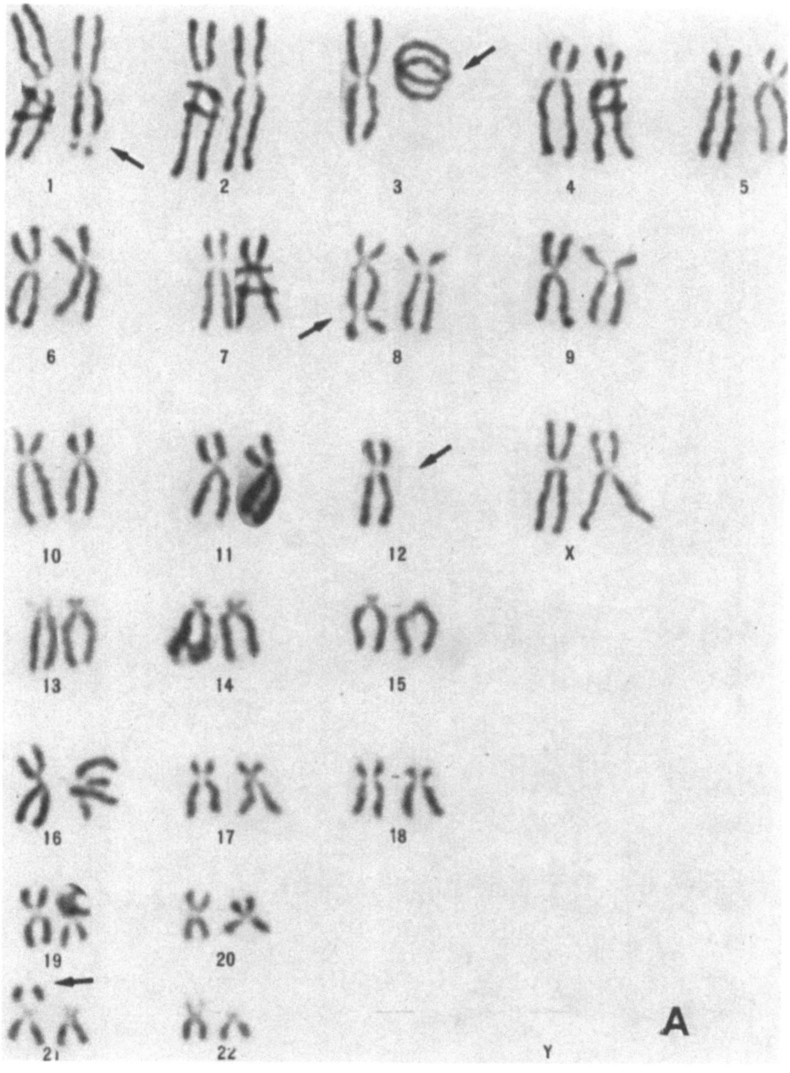

Fig. 5A. 45, XX, C– with onc "ring" chromosome of pairs 3

Fig. 5A–C. Examples of aneuploid mitosis observed in patients

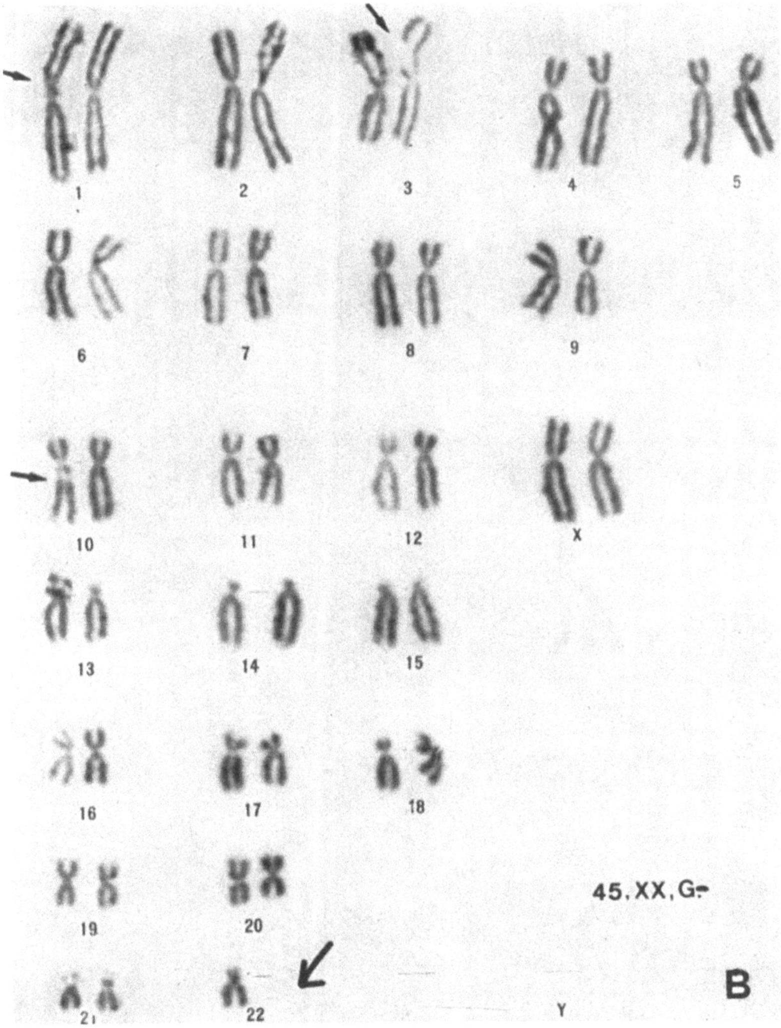

Fig. 5B. 45, XX, G –

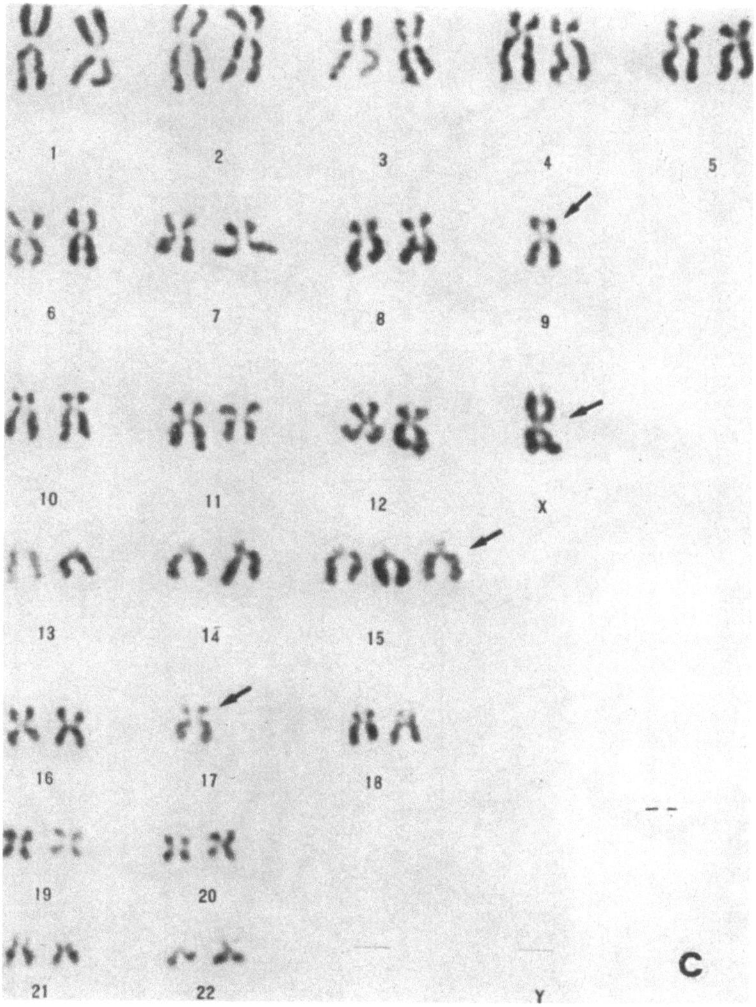

Fig. 5C. 44, XO, C−, E−, D+

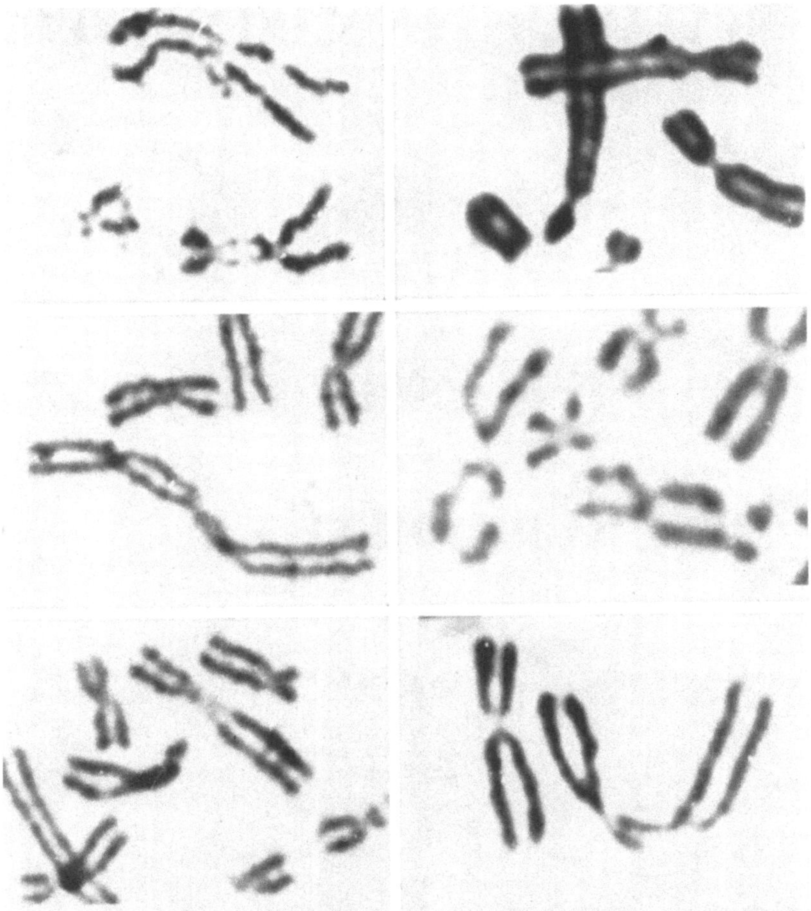

Fig. 5D. Fragmentation and various rearrengements of chromatin observed in one case of BRANDT and SNEDECOR, in its two degrees of freedom. From this statistical analysis only the χ^2 on the data concerning case No. 2 (*T. Claudio*) resulted to be significant (Fig. 6).

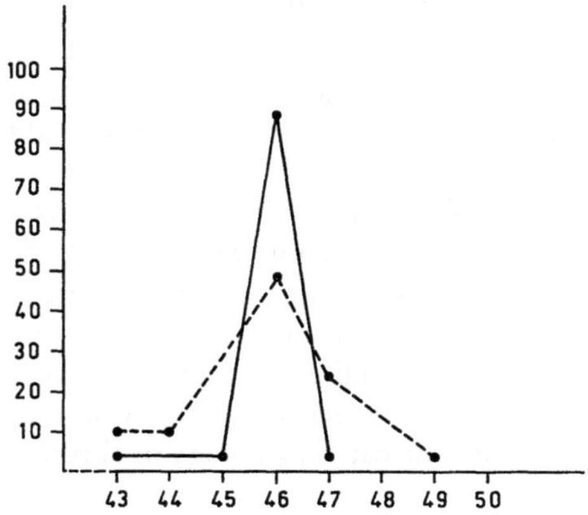

Fig. 6. *T. Claudio*, aged 7 months, suffering from metastatic neuroblastoma. Frequency of aneuploidy before (●—●) and after (● - - - ●) treatment with adriamycin ($\chi^2 = 12.39 - P < 0.001$)

Data obtained by the other subjects were collected in two classes, normal and abnormal cells, and subjected to statistical analysis according to COCHRAN (2).

In all cases, the frequency of abnormal cells was statistically higher after adriamycin treatment ($P = 0.04$), than before treatment. Contrary to what observed in normal human lymphocytes treated "*in vitro*" with adriamycin, chromosomal alterations were present only in one of the patients treated with adriamycin (Fig. 5D). In the lymphocytes collected after the treatment, chromosomal morphology was better than in lymphocytes collected from the same subjects before administration of adriamycin.

Conclusions

Our investigation of the effect of adriamycin on human lymphocytes has shown that these cells are highly affected by the antibiotic.

Adriamycin action on normal human lymphocytes "*in vitro*" is an inhibiting effect of blastogenesis at every concentration tested. The most important finding is that giant cells appear in the lymphocyte cultures treated with low doses of adriamycin. Such cells show a deep disturbance concerning the whole proliferation cycle in its different phases [4]. Also, severe chromosomal lesions were observed at each dose tested; this lead us to formulate the hypothesis that adriamycin can be a mutation-inducing substance. It must be remembered that, in this first experiment, we tested the effects of a direct adriamycin lymphocyte interaction, without any interference by those factors which can be present "*in vivo*", such as drug metabolism, diffusion rate from blood to tissue, special affinity for certain organs, etc. On the contrary, lymphocytes from patients submitted to adriamycin treatment did not show an increase of chromosome aberrations, but a noticeable aneuploidy, higher than the one already existing before treatment. This finding is in accordance with the hypothesis that adriamycin has a mutagenic action.

Summary

The authors have carried out an experimental study in order to investigate the effects of adriamycin on blastogenesis and chromosomes of human lymphocytes. In the first experiment adriamycin (from 0.05—5mcg/ml) was added at 0 h to cultures of normal human lymphocytes.

A constant inhibition of blastogenesis was observed, stronger when concentrations increased. Besides, at lower concentrations many giant cells with a diameter up to 70 micron appeared. Severe chromosome alterations were also present in lymphocytes culture treated with adriamycin concentrations between 0.05 and 0.5 mcg/ml, principally consisting of: fragmentations, rearrangements, chromatid exchanges, polyradial figures, rupture of centromeres with subsequent longitudinal separation of chromatides, alterations of the normal morphology of chromatids, which showed an eroded appearance.

At higher concentrations no mitoses were present.

In the second experiment the lymphocytes of 8 children suffering from advanced malignant tumors, not previously treated with antiblastic drugs or radiations, were studied after "*in vitro*" cultivation. The tests were performed immediately before and the day after the treatment with adriamycin. The drug was given at a daily dose of 0.5 mg/kg, in two courses of three days each, with a three-day interval between them. No inhibition of the ability of lymphocytes to blastic transformation was observed and no giant cells appeared. Increase of aneuploidy was observed, which was statistically significant.

The authors conclude that adriamycin has a mutagenic action as demonstrated in the "*in vitro*" experiments, through the appearance of numerous severe chromosomal changes, and "*in vivo*" through a remarkable aneuploidy. The same action would be responsable for the other effects observed with the "*in vitro*" research, such as inhibition of blastogenesis, appearance of giant cells and polyploidy.

References

1. Arcamone, F., Cassinelli, G., Fantini, G., Grein, A., Orezzi, P., Pol, C., Spalla, C.: Biotech. Bioeng. **6**, 1101—1110 (1969).
2. Cochran, A.: In: Maxwell, A. E.: Analysing qualitative data, p. 77. London: Methuen and Co., 1967.
3. Di Marco, A., Gaetani, M., Scarpinato, B.: Cancer Chemother. Rep. **53**, 33 (1969).
4. Massimo, L., Dagna-Bricarelli, F., Fossati-Guglielmoni, A.: Rev. Europ. Et. Clin. Biol. **7**, 793—799 (1970).
5. Silvestrini, S., Gambarucci, C., Dasdia, T.: Tumori **56**, 137—146 (1970).
6. Vig, B. K.: Cancer Res. **31**, 32—38 (1971).

Action of Adriamycin on HeLa Cells. Evidence of a G_2 Inhibition

D. N. Wheatley

Department of Pathology, University of Aberdeen

With 6 Figures

My interests lie in the basic cellular events which control the progression of a cell through a division cycle, the events which get cells from G_0 into G_1, G_1 into S, S into G_2 and G_2 into M (Fig. 1). For some years we have been concentrating on the

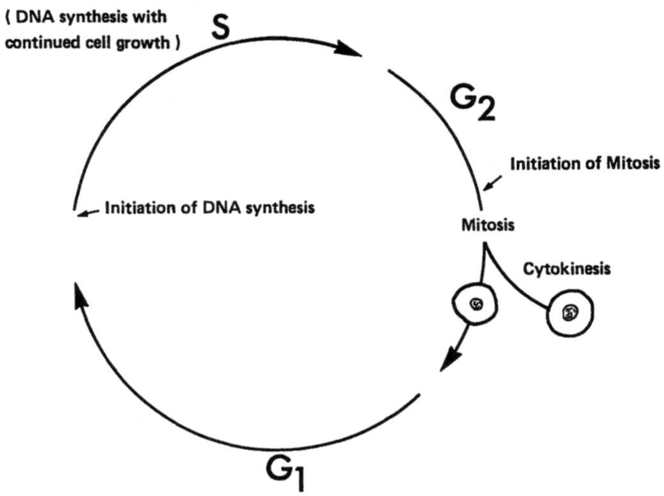

Fig. 1. Schematic diagram of the phase of the cell cycle

last of these, the study of the essential requirements of a post S-phase cell which allow it to enter mitosis, i.e. we are concerned with the significance of the G_2 interval in mammalian cells. Any treatment, physical or chemical, which alters the progression of cells through G_2 and mitosis is potentially of interest in helping to analyse the rate-limiting (control) mechanisms. Adriamycin (Ad) and daunomycin (Da) come into this category.

From the work of DiMarco, Silvestrini and others [1–3] it can be inferred that these antibiotics act within 2 hours of addition to cell cultures (e.g. HeLa cells). Thus the answer to the question of whether they operate antimitotically against G_2 cells is almost certainly that they do. However, we set out first to establish this in an *in vitro* cell culture system, also using HeLa cells, and also to determine more accurately the rate at which their action on the cell cycle occurred.

Table 1. Premises for G_2 studies employing the use of hydroxyurea in the test system

1. Cells will only enter G_2 after completion of 100% of their genomic DNA.
2. The majority of cells in G_2 will normally progress into mitosis.
3. The time taken for cells to progress through G_2 and mitosis is unaffected by the manipulations of the test system.

For this purpose, a hydroxyurea-S phase exclusion system was used, which is based on certain premises given in Table 1 and for which we have good evidence on every point. The most important point is the first, that a cell cannot leave S phase witout completing 100% of its genomic replication. The rationale of the experiment is that on addition of hydroxyurea to a non-synchronously growing cell population, cells in S phase are no longer allowed to progress into G_2 and mitosis but G_2 cells can progress smoothly into mitosis. After an interval of time equivalent to G_2, no further cells come into mitosis and the index in the culture falls to very lowlevels quite abruptly (Fig. 2). We have therefore followed through mitosis a complete cohort of G_2 cells *with no contamination by S phase cells.* There are certain desiderata of

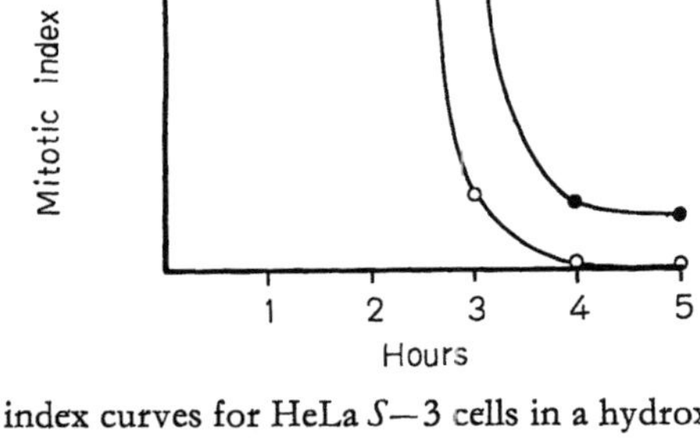

Fig. 2. Typical mitotic index curves for HeLa S–3 cells in a hydroxyurea-S phase exclusion system. ☆-mitotic index of control cultures taken as 100% at every time point. ●-mitotic index of cultures treated with hydroxyurea (2mM) at 0 h showing decline after a period equivalent to the G_2 interval. ○-prophase index which more accurately delineates G_2 as being of between 2 and 3 h duration

the S phase inhibitor in such a system and hydroxyurea has proved to be the agent of choice for the reasons given in Table 2.

Table 2. Hydroxyurea: Suitability for G_2 analysis in cultured cells

At 2mM:

1. DNA synthesis is suppressed by more than 95%.
2. Near maximal inhibition is achieved within a few minutes of addition.
3. RNA and protein synthesis remain unaffected for a period considerably longer than the experimental time.
4. Passage of cells through G_2, M and G_1 is not affected.
5. Almost complete reversibility can be achieved with some cell types (e.g. HeLa) for periods of exposure up to a cell generation less the average S phase time.

We many add to this list one further attribute of hydroxyurea which has been gained from considerable experiments with it over a number of years, namely that we have seen no adverse interaction with any other agent under investigation and therefore it has been successfully used in combination treatments.

We set about testing Ad and Da in this system. Fig. 3 gives the results with Ad at 0.05, 0.1 and 0.5 μg/ml in the presence of hydroxyurea on HeLa cells. Activity is detectable at 0.05 μg/ml but is marked at 0.1 and 0.5 μg/ml. The fall in mitotic index is so rapid in fact that the curves almost correspond to the curves obtained for the progression of a pure culture of mitotic cells into the next interphase. On a comparative basis we could interchange Da for Ad on this graph – i.e. the dose-response curves for Da were identical under the conditions of this experiment.

We may draw several firm conclusions from this study:

1. Firstly, proof is presented that the antibiotics act on G_2 cells.

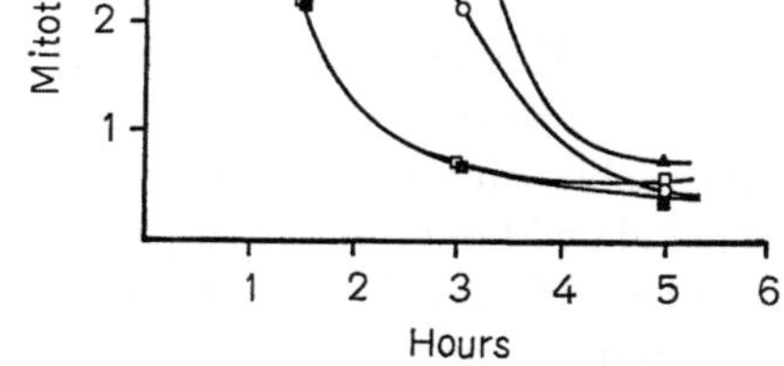

Fig. 3. Effect of adriamycin on G_2 progression in HeLa $S-3$ cells. The mitotic indices are given as actual percentages of cells in mitosis. ● - - ● controls given no treatment; ▲ – ▲ controls given 2mM hydroxyurea alone. ○ – ○ 0.05 μg/ml adriamycin + 2mM hydroxyurea ■ – ■ 0.1 μg/ml adriamycin + 2mM hydroxyurea. □ – □ 0.5 μg/ml adriamycin + 2m hydroxyurea

2. The drug must be taken up in effective concentrations within a very short period of time (Rusconi and Di Marco [2] show near maximal uptake *within* 20 min for Da).

3. The antibiotics do not prevent mitosing cells from progressing into G_1.

4. Low dose levels are required to inhibit mitosis suggesting high specificity of action. 0.1 μg/ml antibiotic is a very effective level of either Ad or Da, and this concentration gives as much inhibition as 0.5 μg/ml – thus the antibiotics probably saturate a limited number of target sites in the cell.

Our studies extended into the effects, at and around the critical 0.1 μg/ml dose level, of Ad and Da on protein synthesis and RNA synthesis in HeLa cultures. Since we are concerned with only the G_2 cohort of cycling cells by excluding S phase cells with hydroxyurea, effects of the antibiotics on DNA synthesis were not investigated. Silvestrini et al. [1] have shown, however, that at 0.1 and 0.5 μg/ml there is only about a 10% inhibition of DNA synthesis over 2 hours. Over a 5 h observation period (with points at 1, 3 and 5 h) we failed to detect any effect of Ad or Da at concentrations up to 0.5 μg/ml on protein synthesis (Fig. 4). This is shown as relative

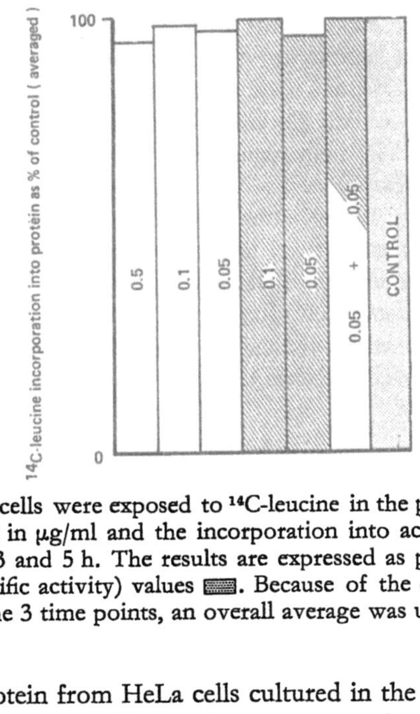

Fig. 4. Cultures of HeLa cells were exposed to ^{14}C-leucine in the presence of adriamycin ▩, daunomycin □ or both ◪ in μg/ml and the incorporation into acid insoluble protein precipitates measured at 1, 3 and 5 h. The results are expressed as percentage of the control cpm/μg protein (i.e. specific activity) values ▦. Because of the consistency of readings at the 3 time points, an overall average was used

specific activities of protein from HeLa cells cultured in the presence of ^{14}C-leucine at different drug concentrations. The results are expressed as the overall averages of the 3 points. The points themselves all fluctuate within a percent or two of the control figures and we can conclude without doubt that protein synthesis is unaffected. In order to determine whether Ad and Da are interchangeable, i.e. have identical biological activities, we began using combinations of the two antibiotics as shown in this Figure. In this same experiment we also used ^{3}H-uridine given as half hour pulses to the same duplicate cultures at each time point and drug concentration. Thus the ^{14}C-leucine served as an internal control in this double labelling experiment. RNA synthesis compared by relative specific activities of the ^{3}H-UdR incorporated did show dose dependent depression (Fig. 5). At each of the time points (1, 3 and 5 h) the suppression of RNA synthesis induced by any particular dose level was of the same degree in relation to the control and therefore the results have been averaged in Fig. 5. From these results it might appear that Da is slightly more effective than Ad in inhibiting RNA synthesis. In terms of anti-mitotic activity there was evidence, however, that the biological effects of equimolar concentrations of the antibiotics were of about the same order and that in combination their effects tended to summate.

To establish this point a more carefully cuoducted experiment was carried out with ^{3}H-UdR at low dose levels of Ad and Da, Fig. 6 shows that Ad and Da had

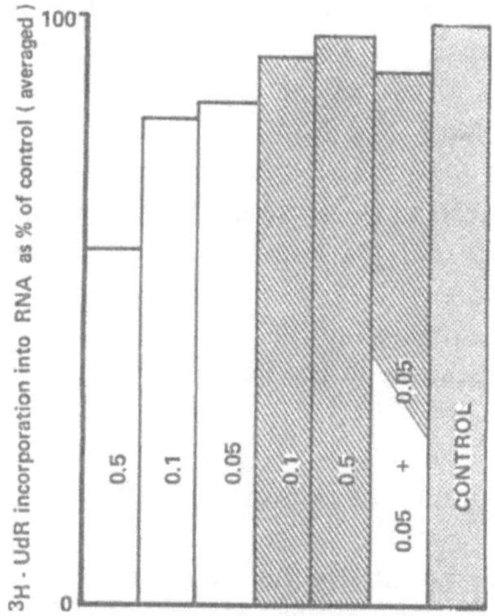

Fig. 5. Effects of the antibiotics on RNA synthesis. ^{3}H-uridine was given as half hour pulses at times of $^1/_2$—1 h, $2^1/_2$—3 h and $4^1/_2$—5 h after exposure of cultures to adriamycin and daunomycin (μg/ml) as in Fig. 4. The results in Fig. 4 serve as an internal control in this double-labelling experiment. Again the specific acticity of tritiated uridine incorporated is expressed as a percentage of the control averaged for the 3 time points

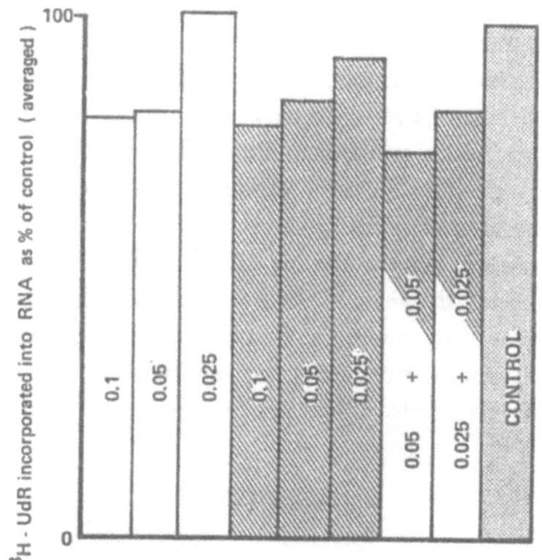

Fig. 6. Examination of adriamycin and daunomycin effects at low concentrations on RNA synthesis, separately and in combination treatments. Details as for Fig. 5

virtually the same RNA inhibitory activity and that at 0.05 and 0.025 μg/ml dose levels given in combination, their actions summated. Whether this same summation effects holds true for the antimitotic effects of Da and Ad is difficult to assess at the very low dose levels since extremely large numbers of cells must be counted before small differences could be verified statistically but preliminary observations suggest that Da is probably somewhat more antimitotic than Ad.

With regard to the interpretation of these results, critical dose levels of Ad and Da which are effective anti-mitotically have, at best, only a slight effect on RNA synthesis. It would be misleading to infer that a cause-and-effect relationship exists between inhibition of RNA synthesis and anti-mitotic activity. Indeed over the past 3 years we have been accruing a wealth of evidence which shows that new RNA synthesis within most, if not all, of the G_2 phase is irrelevant for smooth $G_2 \rightarrow M$ progression in most cell lines in culture. This point has already been discussed in relation to the activity of Da by Silvestrini et al. [3]. It is more likely, therefore, that the two observable effects of the antibiotics are related by a single common cause, e.g. the interaction of Ad and Da with DNA preventing both transcription and chromosome condensation.

Important questions concerning the mechanism of action of these interesting antibiotics and their application in the field of chemotherapy of malignant diseases are still to be answered.

We might ask:

1. What is the basic lesion in the cell caused by Ad and Da which causes inhibition of G_2 progression?

2. Do Ad and Da have essentially the same biological activity?

3. Are all dividing cell populations inhibited by Ad or Da, or do they show discrimination in the types of cells upon which they act, i.e. is there any differential sensitivity?

It is of considerable importance that investigators dealing with the academic problems of cell growth and division and its regulatory mechanisms should become more involved with chemotherapeutic agents. The goal before the researcher and the clinician is much the same, an ability to exert extraneously a control over cellular proliferation mechanisms. A number of agents presently used in chemotherapy or now receiving clinical trials have also proved useful as tools for the analysis of cell cycle events; each might reveal some small facet of the complex mechanism controlling the entry of cells into division. These substances included ICRF 159, bleomycin, phleomycin and now adriamycin and daunomycin. The point I wish to stress here is that these agents are antimitotic largely through their ability to prevent cells progressing from G_2 into mitosis and thus their mode of action is distinctly "radiomimetic" in nature. Bleomycin appears to cause the same type of damage to DNA as irradiation [4] and this raises the issue of whether other agents such as Ad and Da, which are so similar to bleomycin in their biological action, also cause scission of the DNA helix.

References

1. Silvestrini, R., Gambarucci, C., Dasdia, T.: Tumori **56**, 137–148 (1970).
2. Rusconi, A, Di Marco, A.: Cancer Res. **29**, 1507–1511 (1969).
3. Silvestrini, R., Di Marco, A., Dasdia, T.: Cancer Res. **30**, 966–973 (1970).
4. Terasima, T., Yasukawa, M., Umezawa, H.: Gann **61**, 513–516 (1970).

Part III
Activity against Experimental Tumors

Adriamycin: The Therapeutic Activity on Experimental Tumors

A. DI MARCO

Istituto Nazionale per lo Studio e la Cura dei Tumori
Divisione Oncologia Sperimentale B, Via Venezian, 1-Milano

With 7 Figures

Adriamycin is a new antibiotic isolated from the cultures of a mutant strain of *S. Peucetius*, called. *S. Peucetius var. caesius* and shown to be highly active in inhibiting the growth of Ehrlich ascites tumor (ARCAMONE, CASSINELLI, DI MARCO, and GAETANI, 1969).

Its antitumor activity was compared to that of daunomycin on different experimental tumors and shown to be stronger than that of daunomycin (DI MARCO et al., 1969). As the chemical work of ARCAMONE and coworkers has shown the strict structural relationships with daunomycin (daunorubicin) (ARCAMONE, CASSINELLI, FRANCESCHI, OREZZI, PENCO, and SELVA, 1971, Symposium Internazionale Adriamicina), the question has been raised of the biological basis of the different activity. The right answer to this question could give some hints concerning the broad problem of the desirable properties of an antineoplastic drug.

We will summarize here some of the earlier reports of our laboratory and some more recent ones that seem to be pertinents to the question raised.

In mice transplanted with Ehrlich ascites tumor, adriamycin induces an increase of survival time longer than daunomycin (Table 1) (DI MARCO et al., 1969); it exerts a highly inhibitory effect on the growth of many other experimental tumors (DI MARCO et al., 1969) and on the L 1210 leukemia (SANDBERG et al., 1970).

Table 1. Survival time ratio of mice with Ehrlich ascites tumor treated i.p. for 5 days

Dose (mg/kg/day)	Survival time in days[a]			
	Daunomycin		Adriamycin	
	T/C	Ratio	T/C	Ratio
2,50	29.5/16.4	1.8	43.9/15.7	2.8

[a] Each value is the average of the results obtained in 10 animals per group.

The effect of adriamycin on the growth of sarcoma 180, transplanted subcutaneously in mice, is shown in Table 2. It can be seen that the antibiotic markedly inhibited tumor growth at each dose used. A significant mortality of the treated animals was noted only with the highest dosage (7 mg/kg/day).

Table 2. Effect of adriamycin on mice with Sarcoma 180

Treatment dose (mg/kg/day) and route	Body weight change (g)	Mean tumor weight (g)	Percent inhibition	Mortality
Controls[a]	+ 4.78	3.922 ± 1.012		1/10
Controls[b]	+ 5.65	2.469 ± 0.852	—	0/10
Adriamycin				
7 × 8 sc[a]	− 5.98	0.239 ± 0.097	93.9	6/10
5 × 8 sc[b]	− 3.37	0.241 ± 0.088	90.2	0/10
3.50 × 8 sc[a]	− 2.31	0.696 ± 0.276	82.3	0/10
2.50 × 8 sc[b]	− 1.60	0.657 ± 0.259	73.4	0/10
1.75 × 8 sc[a]	+ 3.09	1.988 ± 0.647	49.3	0/10

[a] 1st experiment.
[b] 2nd experiment.

Mice bearing sarcoma 180 were treated subcutaneously for 8 days starting 1 day after tumor transplantation. The antibiotic was administered in saline solution at different concentrations.

For therapeutical purposes it is important to compare the antitumor activity of adriamycin and daunomycin in a quantitative manner. For this purpose we used the sarcoma 180 tumor system in mice. The results of this experiment are reported on Table 3.

Table 3. Comparison of the effect of daunomycin and adriamycin on mice with Sarcoma 180

Treatment dose (mg/kg/day) and route	Body weight change (g)	Mean tumor weight (g)	Percent inhibition	Mortality
Controls	+ 4.5	3.443 ± 1.048		0/8
Daunomycin				
2.50 × 8 sc	+ 3.5	2.621 ± 0.863	23.9	0/8
5.00 × 8 sc	+ 2.1	1.752 ± 0.529	49.2	0/8
Adriamycin				
2.50 × 8 sc	+ 2.0	1.703 ± 0.512	50.6	0/8
5.00 × 8 sc	− 0.2	0.672 ± 0.289	80.5	0/8

Adriamycin clearly had a higher activity than daunomycin on this tumor. As yet published (Di Marco et al., 1969) the ID 50 in sarcoma 180 was about 3.3 mg/kg for daunomycin and about 1.5 mg/kg for adriamycin. The data are confirmed by the results shown in Table 3 in which the antibiotics were administered subcutaneously in doses of 2.5 or 5 mg/kg/day for 8 days. Statistical analysis of these results, performed by the multiple range method, has further demonstrated that both daunomycin and adriamycin significantly inhibit tumor growth but the effectiveness of adriamycin is about twice that of daunomycin. Moreover, by plotting the dosage-mortality and dosage-response data reported in Table 4 and 2, it is possible to calculate, according to Skipper and Schmidt, the therapeutic index of the antibiotic (Table 5). Under the same experimental conditions the therapeutic index for daunomycin is 0.67.

Table 4. Subacute toxicity of adriamycin in albino mice treated sc for 8 days

Dose (mg/kg/day)	Mortality at 10th day Number	%	15th day Number	%
10	10/10	100	10/10	100
8.33	7/10	70	10/10	100
6.67	4/10	40	8/10	80
5	0/10	0	0/10	0
2.50	0/10	0	0/10	0
1.25	0/10	0	0/10	0

Table 5

$$\mathrm{TI} = \frac{\text{Maximum tolerated dose (LD 10)}}{\text{Minimum effective dose (T/C 0.10)}} = \frac{5.8}{4.8} = 1.21$$

As observed in our laboratory (Di Marco, Lenaz, Casazza and Scarpinato, to be published) adriamycin is active on spontaneous mammary carcinoma of C3H mice, which is of particular interest because this is an autochtonous tumor, highly resistant to most chemotherapeutic agents. Experiments on the best treatment schedule (Table 6) were performed on this tumor transplanted in C3H inbred mice, thus minimizing differences in tumor growth in different hosts due to immunological factors. It has been shown (Scholler, Philips, Sternberg, Bittner, 1956) that tumor characteristics, including susceptibility to chemotherapeutic agents do not vary significantly after a small number of transplants. We therefore used tumors transplanted up to the sixth generation.

Treatment, started the day after tumor implant (Table 7) and repeated every second day with adriamycin 2 mg/kg/day for two subsequent cycles, strongly inhibits tumor growth and produces a highly significant increase in survival time with a high percentage of long term survivors. Treatment with the same dose on consecutive days caused simply a delay in tumor development without any increase in survival time over control value and an increase in metastatic dissemination. These

Table 6. Experimental design

Mammary Carcinoma	Experiment	Drug	Dose (mg/kg/die)	Treatment schedule	Table (T) and figures (F) of results
Spontaneous	1	D	3.25	qd; D 1–6; 12–17	T2 F1
		A	2.0–2.25–2.5		
	2	D	3.25	q2d; D 1–11; 17–27	T2 F1
		A	2.25		
Transplanted	3	D	2.6–3.25	qd; D 1–8	T3
		A	2.0–2.5		
	4	D	2.6–3.25	qd; D 10–17	T4 F2
		A	2.0–2.5		
	5	D	2.6–3.25	qd; D 1–6; 12–14; 17–19	T6 F3
		A	2.0–2.5		
	6	D	2.6–3.25	q2d; D 1–11; 17–27	T7 F4
		A	2.0–2.5		

Table 7. Daunomycin and adriamycin activity on transplanted mammary carcinoma

Treatment[a]		Males			Females			Total			
Drug	mg/kg/day	Average survival time[b] (days)	tumor bearing animals[d]	Survivors[c]	Average survival time (days)	tumor bearing animals	Survivors	Average survival time (days)	% mals	tumor bearing animals	Survivors
—	—	81.0	5/5	0/5	100	5/5	0/5	90.5	100	10/10	0/10
D	2.6	156.8	5/5	0/5	132.8	5/5	0/5	144.8	160	10/10	0/10
D	3.25	137.8	5/5	0/5	132.2	5/5	0/5	135.0	149	10/10	0/10
A	2.0	180.0	2/5	3/5	169.5	2/5	3/5	174.7	193	4/10	6/10
A	2.5	200.0	1/5	4/5	181.0	1/5	4/5	190.5	210	2/10	8/10

[a] Mice were treated every other day from day 1–11 and from day 17–27 after the transplant.
[b] 210 days after the beginning of the experiment.
[c] Long term survivors are excluded.
[d] Transplanted tumor bearing animals.

negative results can be explained, as will be discussed further, by the cumulative toxicity of adriamycin when administered daily. The frequent increase in metastatic dissemination (to lungs, spleen, liver, kidneys and testicles) observed, could be explained by a suppression, in these conditions, of immunological defence mechanisms (Di Marco, Lenaz, Scarpinato, 1971).

In murine sarcoma virus (Moloney) induced tumors (Fig. 1) in 16-day-old Swiss CD-1 mice, daunomycin causes a reduction of tumor growth rate followed by recurrence of tumors which progress until they kill the mice (Casazza et al., 1970). Untreated control mice show an initial equivalent tumor growth but then undergo complete and permanent tumor regression. Tumor recurrence is rare after daunomycin treatment in 20-day-old mice and is absent after daunomycin treatment in 28-day-old mice at the time of infection. Tumor recurrence can be attributed to the action of daunomycin on the induction and multiplication of immunocompetent

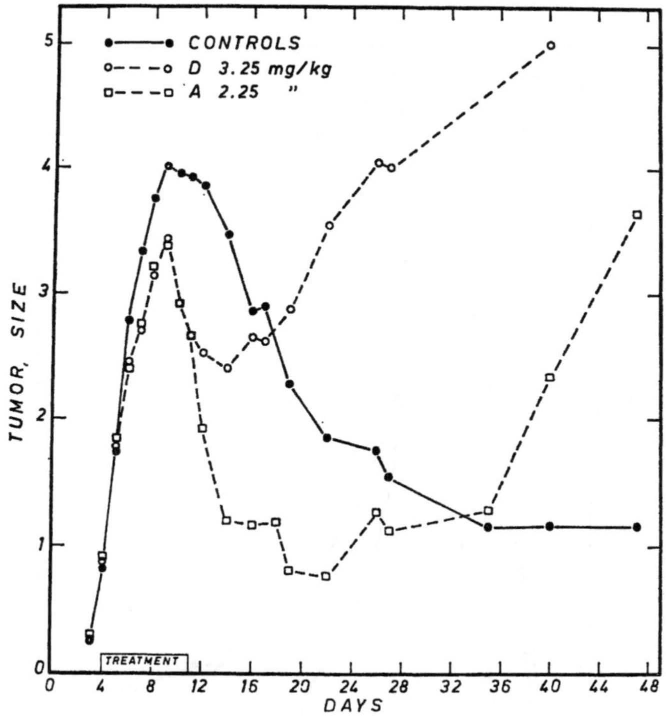

Fig. 1. Effects of daunomycin and adriamycin, administered on days 3→10, on (MSV) M-induced tumor

cells which ordinarily react, directly or indirectly, with the virus-induced antigens in tumor cells. Thus the effect of the chemotherapy on the host adversely effects the complex interrelationships between host, tumor and virus, which would have led to tumor regression.

The dependence of tumor recurrence on animal age has tentatively been explained by considering the development of immunologically competent cells in early life, and the high activity exerted by daunomycin on dividing cells. Daunomycin-treated animals show diminished lymph nodes and spleen weight; the younger the animals the more susceptible their lymphoid cells are to the antibiotic.

When compared to daunomycin, adriamycin was more effective in inhibiting tumor growth; tumor recurrence occurred later, and to a lesser extent. When administered before the infection (Fig. 2) daunomycin inhibited the probable tumor regression, whereas adriamycin did not adversely affect the natural regression. In these experimental conditions the difference between the two antibiotics is particularly evident.

We are now faced with the question of the reasons for the different sensitivity to daunomycin and adriamycin of induction and multiplication processes of immunocompetent cells. But we have no definite answer to this question and it is only very tentatively that we may think of a differential access of the drugs to the place where immunocompetent cells are formed.

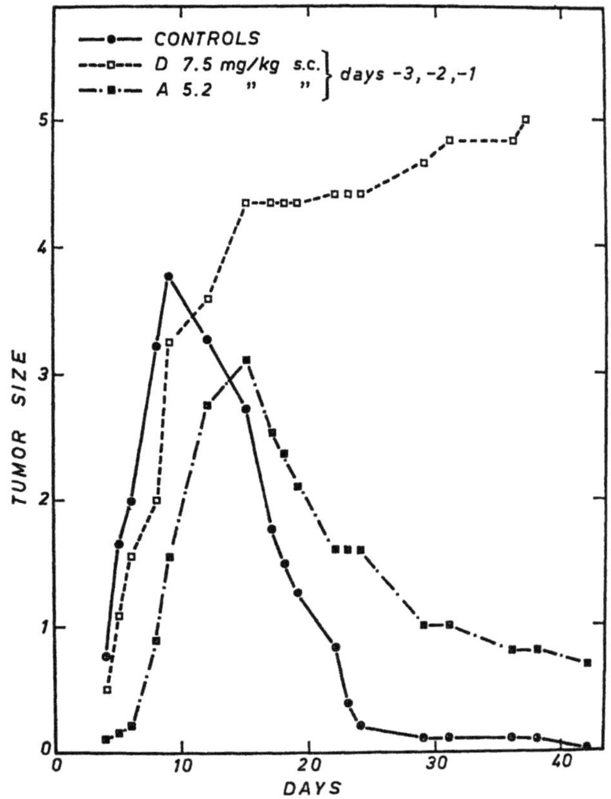

Fig. 2. Effect of pretreatment with daunomycin and adriamycin on MSV (M)-induced tumor

The higher antitumor activity of adriamycin, as compared to daunomycin, could be due to:

1. higher activity against tumor cells;
2. different pharmacodynamic behaviour.

1. The in vitro activity of both drugs was studied on different cell lines (NECCO, DASDIA, 1971, unpublished results). In the HeLa cells cultivated *in vitro*, adriamycin has a lower effect on the mitotic index (ID 50, after 8 h treatment for daunomycin = 0.029 μg/ml and adriamycin = 0.14 μg/ml) and on the proliferative activity (ID 50 after 8 h treatment and 40 h recovery, for daunomycin = 0.3 μg/ml and adriamycin = 0.64 μg/ml) than daunomycin.

The lack of parallelism between the dose-effect lines of the two antibiotics does not permit, however, to establish a precise ratio between the two.

Also on different cell lines adriamycin shows a slightly lower effect than daunomycin (Table 8).

Table 8. Percent of reduction in mitotic index, and proliferative activity of different cell lines.

	HELA		PF		JLV6		JLV8		MS	
	M.I.	P.A.	M.I.	P.A.	M.I.	P.A.	M.I.	P.A.	M.I.	P.A.
Adriamycin										
1 μg/ml	100	55.86	95.56	54.94	100	44.07	40.75	60.48	110	31.02
0.1 μg/ml	28.82	7.48	93.34	41.02	100	8.48	74.08	50.00	90.55	26.62
0.01 μg/ml	15.26	5.00	11.12	27.36	100	5.09	25.93	7.15	91.90	0.31
Daunomycin										
1 μg/ml	100	75.64	95.56	63.05	100	44.92	87.66	56.43	100	70.00
0.1 μg/ml	100	12.15	93.34	20.99	92.69	23.73	87.66	48.58	100	21.08
0.01 μg/ml	22.31	9.00	66.67	+2.37	75.61	11.87	61.73	1.43	87.84	15.64

P.F. = human kidney, cell strain.
JLV6 = normal lymphoid cells from Balb/c, cell strain.
JLV8 = JLV6 in vitro infected with Moloney Leukemia Virus, cell strain.
MS = Moloney Sarcoma Virus induced in Balb/c, cell line at 12th passage.

By the appearance of fluorescence in HeLa cells, after administration of the two drugs, is possible to see that adriamycin, like daunomycin, is quickly taken up by the cells and fixed to the nucleolar and perinucleolar aereas. Tritium-labelled adriamycin quickly penetrats into the cells and is fixed to the nuclear structures with a marked localization at the level of perinucleolar chromatine (Silvestrini, Gambarucci, Dasdia, 1970).

The effect of adriamycin on DNA synthesis, as shown by the incorporation of ^{3}H-thymidine in the nucleus by the autoradiography method, seems also to be similar to that of daunomycin (Di Marco et al., 1971).

From the work done on the mechanism of action of daunomycin it is known that the most important property of this molecule is the ability to bind specifically with DNA. All the biological activities of daunomycin and its analogues – inhibitory effect on DNA replication and on DNA-dependent RNA synthesis, antimitotic effect, and ability to induce chromosomal aberrations – are stricly related to this property (Di Marco et al., 1967).

Prèliminary work on the interactions of some daunomycin analogues with DNA (Di Marco et al., 1971) showed that the affinity for DNA of adriamycin, as judged by the ability to displace methyl green from its DNA complex is only slightly lower than that of daunomycin. (Molar ratio antibiotic: DNA P for 50% displacement of methyl green from its complex with calf thymus DNA, daunomycin 0.076, adriamycin 0.09). Also the protection against thermal denaturation of DNA is similar for both drugs; from the DNA intrinsic viscosity increase it can be deduced that adriamycin, as daunomycin is also intercalating between the DNA double-helical structure (Zunino, Gambetta, Di Marco, to be published).

These as yet preliminary data seem to justify the view that adriamycin has a mechanism of action very similar to that of daunomycin. From the cell lines studied in vitro, it also appears that adriamycin does not display a generally higher anti-proliferative activity than daunomycin.

2. The distribution of adriamycin has been studied in normal and sarcoma 180-bearing mice using tritiated drug (Di Fronzo, Gambetta, Lenaz, 1971). After an intravenous injection the radioactivity is rapidly cleared from the blood and taken up into the different organs (Fig. 3). The pattern of distribution of the antibiotic

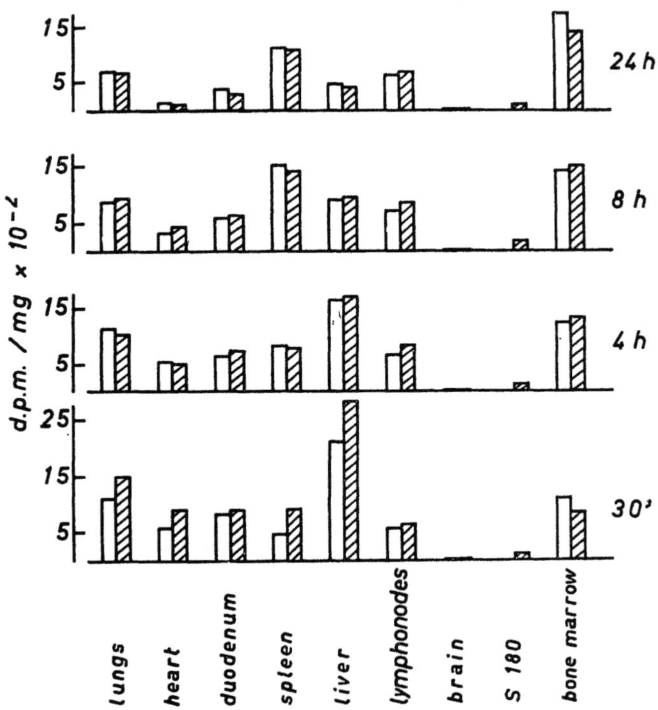

Fig. 3. Distribution of ^{3}H-adriamycin after a single i.v. dose of 5 mg/kg in normal (□) and S 180-bearing (▩) mice. Radioactivity levels for bone marrow are expressed as d.p.m./μg DNA

shows initial high levels and a rapid decrease in most tissues, while in spleen the radioactivity increases slowly and constant levels were maintained for longer periods of time. Low levels were detected in the brain, thus suggesting that neither antibiotics crosses the blood-brain barrier.

Radioactivity levels (Fig. 4) in the different tissues are constantly higher after adriamycin administration. A 50% drug excretion (Fig. 5) occurs in 24 h for daunomycin and in 32 h for adriamycin. Adriamycin appears to be accumulated (Fig. 6) after repeated treatments to a much greater extent than daunomycin. Maximal drug levels were observed 24 h after the third drug administration, a fourth dose did not result in any further increase in radioactivity levels. No evidence of biological transformation has been obtained for adriamycin, while daunomycin appears to be extensively metabolized. In Fig. 7 we have the plot of the function $\log X = \log X_0 - \frac{Kt}{2.303}$ where X is the residual percentage of radioactivity at t time, and X_0 is 100% of total radioactivity administered, after a single injection of 5 mg/kg of daunomycin or adriamycin. The first part of the daunomycin plot corresponds probably to the trans-

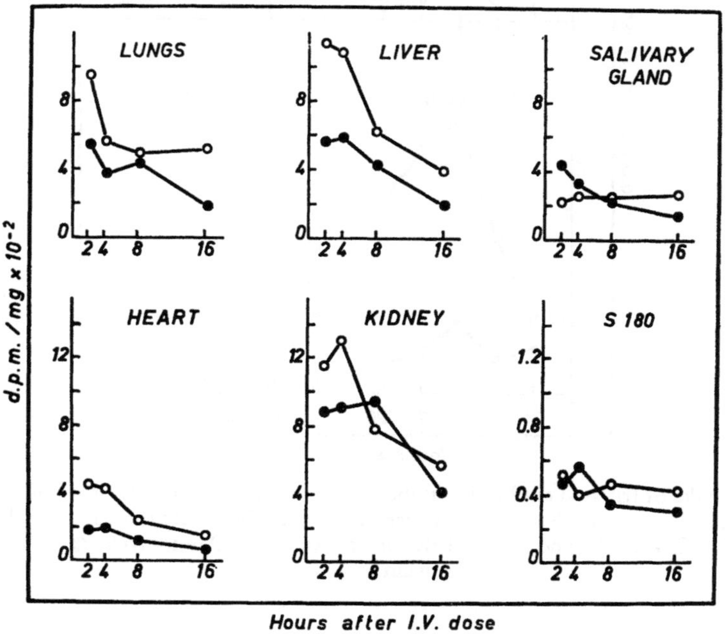

Fig. 4. Radioactivity levels at different times after single i.v. doses of 5 mg/kg of ^{3}H-daunomycin (●) or ^{3}H-adriamycin (○) in lungs, liver, submaxillary salivary glands, S 180, kidney

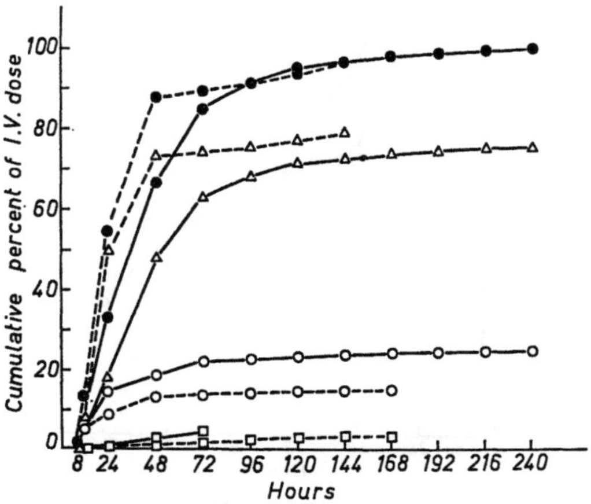

Fig. 5. Radioactivity levels in the urines, (○), faeces (△) and expired air (□) and total excretion (●) expressed as % of the total radioactivity administered i.v. Daunomycin 5 mg/kg (- - - -) and adriamycin 5 mg/kg (———)

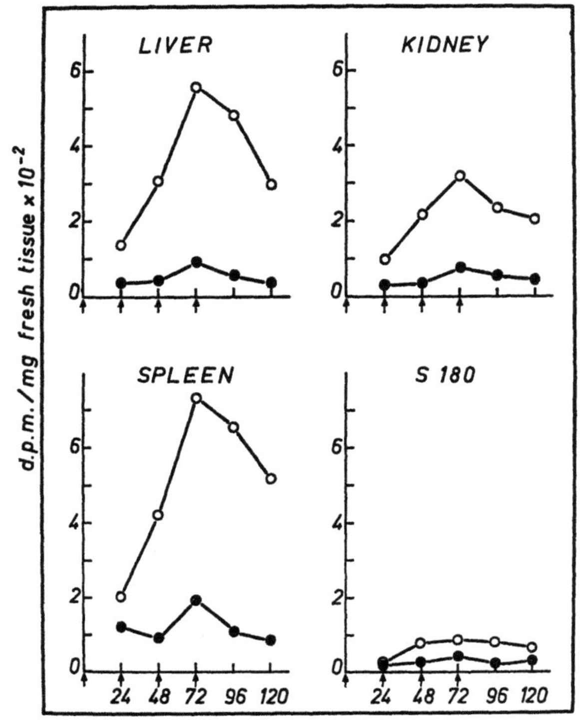

Fig. 6. Uptake of tritiated antibiotics in the tissues of mice given daily doses of 3 mg/kg/day i.v. for 1, 2, 3 or 4 days. Radioactivity was determined 24 h after the end of each treatment cycle and 48 h after the end of the 4 day treatment cycle. Daunomycin ●——●, Adriamycin ○——○

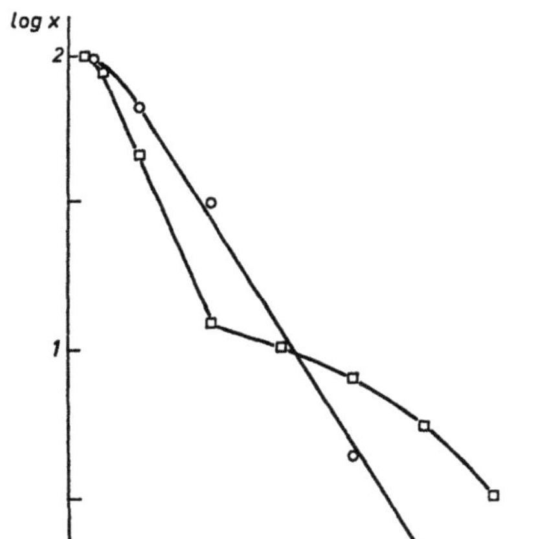

Fig. 7. Plot of the function $\log X = \log X_0 - \frac{Kt}{2.303}$ where X is the residual percentage of radioactivity at t time, and X_0 is 100% of total radioactivity administered, after a single injection of 5 mg/kg of daunomycin (○) and adriamycin (□)

formation of daunomycin in its metabolic product daunorubicinol (DI MARCO et al., 1967; BACHUR, 1971). These findings explain the higher cumulative toxicity of adriamycin versus daunomycin and they could also be consistent if consideration is given to the factor concentration per time.

Concluding these remarks, the more favourable therapeutic ratio of adriamycin, as compared to daunomycin, is probably not related to an higher activity in tumor cells or to a different mechanism of action. The pharmacokinetic behaviour possibly resulting from a different metabolism of the drugs should be considered responsible for the better therapeutic activity.

References

ARCAMONE, F., CASSINELLI, G., DI MARCO, A., GAETANI, M.: Patent Application, Farmitalia Research Laboratories, 1969.

ARCAMONE, F., CASSINELLI, G., FRANCESCHI, G., OREZZI, P., PENCO, S., SELVA, A.: La struttura della adriamicina nuovo antibiotico antitumorale. Symposium Internazionale Adriamicina, 71° Convegno Farmitalia, 1971.

BACHUR, N. R.: J. Pharmacol. exp. Ther. **177**, 573 (1971).

CASAZZA, A. M., SCARPINATO, B. M., DI CUONZO, G.: Activity of adriamycin and daunomycin on primary and transplantable Moloney Sarcoma Virus-induced tumors in mice. 10th International Cancer Congress, Houston, May 22—29, 1970. Abstracts, pp. 391, 1970.

DI FRONZO, G., GAMBETTA, R. A., LENAZ, L.: Rev. Europ. Etud. Clin. Biol. **16**, 572 (1971).

DI MARCO, A.: Daunomycin and related antibiotics. Antibiotics-Vol. 1, Mechanism of action, 1967.

DI MARCO, A., BORETTI, G., RUSCONI, A.: Il farmaco, **22**, 535 (1967).

DI MARCO, A., GAETANI, M., SCARPINATO, B. M.: Cancer Chemother., Rep. **53**, 33 (1969).

DI MARCO, A., LENAZ, L.: Daunomycin and adriamycin. Cancer Med. (in press).

DI MARCO, A., LENAZ, L., CASAZZA, A. M., SCARPINATO, B. M.: Activity of adriamycin (NSC-123, 127) and daunomycin (NSC-82151) on mouse mammary carcinoma. Cancer Chemother. Rep. (in press).

DI MARCO, A., LENAZ, L., SCARPINATO, B. M.: Protezione immunologica legata al sesso nel carcinoma mammario murino. Congresso di Cancerologia, Cagliari 8—9 Ottobre 1971.

DI MARCO, A., ZUNINO, F., GAMBETTA, R. A.: Unpublished data.

DI MARCO, A., ZUNINO, F., SILVESTRINI, R., GAMBARUCCI, C., GAMBETTA, R. A.: Biochem. Pharmacol. **20**, 1323 (1971).

NECCO, A., DASDIA, T.: Unpublished data, 1971.

SANDBERG, S. J., HOWSDEN, F. L., DI MARCO, A., GOLDIN, A.: Cancer Chemother. Rep. **51**, 1 (1970).

SCHOLLER, J., PHILIPS, F. S., STERNBERG, S. S., BITTNER, J. J.: Cancer **9**, 240 (1956).

SILVESTRINI, R., GAMBARUCCI, C., DASDIA, T.: Attività biologica dell'adriamicina "in vitro". Tumori **56**, Nr. 3 (1970).

SKIPPER, H. E., SCHMIDT, L. H.: Cancer Chemother. Rep. **17**, 179 (1962).

Some Factors Influencing the Chemotherapeutic Effectiveness of Adriamycin

Abraham Goldin

Drug Research and Development, Chemotherapy, NCI, Bethesda, Maryland, USA

With 3 Figures

A variety of factors may influence the antitumor effectiveness of drugs and if they are appropriately manipulated may lead to improved therapeutic effectiveness.

Structural modification of an antitumor agent may lead to improved therapeutic activity. This has been amply illustrated in the leukemia L 1210 system which has been used so extensively for the identification and evaluation of the comparative therapeutic efficacy of antitumor agents in the Chemotherapy program of the National Cancer Institute in the United States. Adriamycin, the new antitumor-antibiotic obtained from cultures of *Streptomyces peucetius var. caesius* [1] differs from the structural analog daunomycin only in the substitution of a hydroxyl group for hydrogen in the acetyl radical of the aglycone moiety of daunomycin [2]. This minor structural modification has resulted in improved therapeutic effectiveness for adriamycin over daunomycin in a spectrum of animal tumor systems. Its therapeutic index was higher than that of daunomycin in the treatment of sarcoma 180, Ehrlich ascites carcinoma, Oberling-Guérin-Guérin carcinoma, and a nitrosomethylurea induced lymphosarcoma of C3Hf mice [3].

Adriamycin is more effective than daunomycin in the leukemia L 1210 system in mice. This is illustrated in Fig. 1 where with treatment on a single day (day 1) or with daily treatment (days 1–9), over a series of dose levels adriamycin was more effective than daunomycin in increasing the survival time of the leukemic mice. This experiment illustrates a primary factor which can influence drug effectiveness, namely the toxicity of the drug for the host. For both adriamycin and daunomycin, as with other drugs that have antitumor activity, as the dosage of the drug is increased, the survival time of the animals is increased. At an optimal dosage, where the toxicity for the host is equivalent to the antileukemic effectiveness, the maximum increase in survival time is achieved. With further increase in dosage the toxicity of the drug for the host becomes dominant and the survival time diminishes. This emphasizes that in chemotherapy there is a balance between the effectiveness of a drug against the tumor cells and the toxic action of the drug on the host. This relationship underlies the specificity of action of a drug and there is abundant evidence that this specificity can be altered in favor of increased antitumor effectiveness. It is possible to increase the toxicity of the drug for the tumor without a proportionate increase in toxicity for the host. Conversely it is possible to diminish the toxicity of a drug for the host without a proportionate loss in antitumor effectiveness. In either event there will result an increase in therapeutic specificity of drug action. The host-tumor-drug interrelationships are thus of paramount importance and their interplay determines the outcome of chemotherapy.

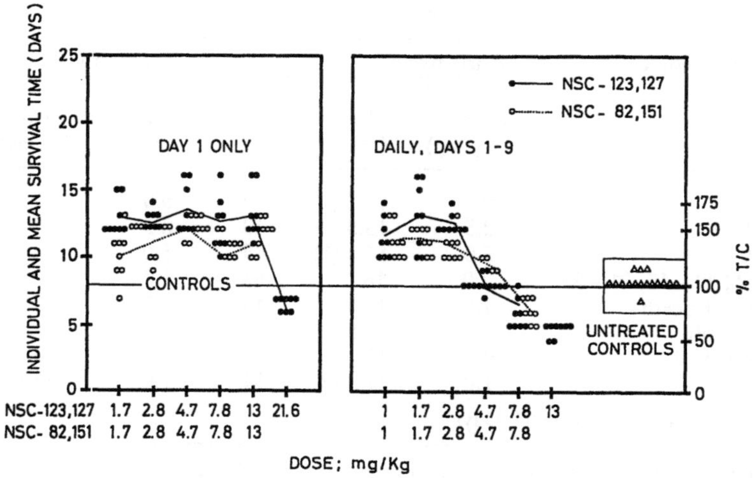

Fig. 1. Comparison of adriamycin (NSC−123,127) and daunomycin (NSC−82,151) against early leukemia L1210. Data of Kline et al., Microbiological Associates, Inc.

The data obtained to date with adriamycin apply primarily to the manipulation of the effectiveness of the drug on the tumor cells relative to the toxic effect on the host by alteration of the scheduling of therapy. Also entering into consideration would appear to be the relative rates of recovery of the tumor cell population and the host from drug toxicity. There has been considerable experience in experimental tumor systems showing that the schedule of therapy may have a profound effect in many instances on the therapeutic outcome. The initial example involved a scheduling study with methotrexate in the treatment of leukemia L 1210. It was demonstrated that treatment with methotrexate every four days in early stages of leukemia was considerably more effective than daily treatment in increasing the survival time of the mice [4]. In parallel manner it was found that intermittent treatment with methotrexate provided a marked extension of remission maintenance in children with acute lymphocytic leukemia [5]. These primary observations have resulted in considerable emphasis on the importance of scheduling and in our chemotherapy program all of the drugs of interest are being evaluated for their schedule dependency characteristics [6].

Similarly adriamycin is also schedule dependent and it is of interest to analyze the nature of this schedule dependency. The therapeutic advantage of adriamycin over daunomycin for a wide range of dosage schedules in the treatment of leukemia L 1210 is illustrated in two experiments (Tables 1, 2) [7]. Both adriamycin and daunomycin increased the survival time of mice with leukemia L 1210 over a wide range of treatment schedules. But for all of the treatment schedules in the two experiments, the greater effectiveness of adriamycin over daunomycin in prolonging life span was evident. Also, adriamycin produced a somewhat greater number of long term survivors than did daunomycin. The marked advantage of an every 3 h schedule (8 injections a day) for a single day for adriamycin is evident in each of the two experiments.

Table 1. Effect of schedule of therapy on the antileukemic (L1210) effectiveness of adriamycin and daunomycin[a] [7]

Treatment schedule	Adriamycin			Daunomycin		
	Optimal dose (mg/kg)	MST (Days)[b]	ILS (%)[c]	Optimal dose (mg/kg)	MST (Days)	ILS (%)
Day 1 only	10.0	15.5 (1)	40	5.0	13	18
Days 1, 5, 9	2.5	15.5	40	2.5	14	27
Days 1, 9	5.0	16 (1)	45	5.0	14	27
Days 1–5, 11–15, 21–25	2.0	18	63	1.0	14.5	31
Daily, days 1-death	1.0	15.5	40	0.5	14	27
q3h day 1 only	1.0	32.5 (3)	195	1.0	16 (1)	45
q3h days 1, 5, 9	0.5	15	36	0.5	14	27
q3h days 1, 9	1.0	17 (2)	54	1.0	15.5	40
Untreated controls MST = 11 Days						

[a] BDF_1 male mice (18–26 g) inocluted I.P. (day 0) with 10^5 L1210 cells taken from DBA/2 mice bearing ascitic L1210. For each treatment schedule the drugs were administered I.P. over a series of dosage levels. q3h = every 3 h; optimal dose is per injection.
[b] MST = Median survival time. The number of long term survivors are given in parenthesis. ( ) = Survivors.
[c] ILS (%) = Percent increase in median survival time over controls.

Table 2. Effect of schedule of therapy on the antileukemic (L1210) effectiveness of adriamycin and daunomycin[a] [7]

Treatment schedule	Adriamycin			Daunomycin		
	Optimal Dose (mg/kg)	MST (Days)[b]	ILS (%)[c]	Optimal dose (mg/kg)	MST (Days)	ILS (%)
Day 1 only	10.0	13	62	2.5	12	50
Day 1, 5, 9	5.0	13.5	68	5.0	12.5	56
Daily, Day 1–7	2.0	13	62	2.0	12	50
Day 1–5, 11–15, 21–25	2.0	13	62	1.0	12.5	56
q3h day 1 only	2.0	16.5	106	1.0	13	62
q3h days 1, 3	0.5	14 (1)	74	0.5	13	62
q3h days 1, 3, 5	0.5	13.5	68	10.0	13	62
q3h days 1, 5, 9	1.0	13.5	68	0.5	13	62
Untreated controls MST = 8 days						

[a] BDF_1 male mice (18–26 g) inoculated I.P. (day 0) with 10^5 L1210 cells taken from DBA/2 mice bearing ascites L1210. For each treatment schedule the drugs were administered I.P. over a series of dosage levels. q3h = every 3 h; optimal dose is per injection.
[b] MST = Median survival time. The number of long term survivors are given in parenthesis. ( ) = Survivors.
[c] ILS (%) = Percent increase in median survival time over controls.

With daunomycin the treatment schedule of every 3 h for a single day resulted in only slightly greater increases in the survival time as compared for example with daily therapy.

Table 3. Influence of schedule of therapy and number of treatments per day on the antileukemic (L1210) effectiveness of adriamycin[a]

Days of treatment	Exp. No.	Number of treatments per day: Single — Optimal dose (mg/kg)	Single — ILS (%)[c]	Eight (q3h) — Optimal dose (mg/kg)	Eight (q3h) — ILS (%)
Day 1 only	1	10.0	40 (1)	1.0	195 (3)
	2	10.0	62	2.0	106
	3[b]			1.8	111 (2)
Days 1, 3	2			0.5	74 (1)
Days 1, 3, 5	2			0.5	68
Days 1, 5, 9	1	2.5	40	0.5	36
	2	5.0	68	1.0	68
Days 1, 9	1	5.0	45 (1)[d]	1.0	54 (2)
	3			1.8	111
Days 1, 11	3			1.8	78
Days 1, 13	3			1.8	95
Days 1, 15	3			1.8	100
Day 1–7	2	2.0	62		
Day 1-death	1	1.0	40		
5 days on,	1	2.0	63		
5 days off (Days 1–5, 11–15, 21–25)	2	2.0	62		

[a] BDF_1 male mice (18–26 g) inoculated I.P. (day 0) with 10^5 L1210 cells taken from DBA/2 mice bearing ascites L1210. For each treatment schedule the drugs were administered I.P. over a series of dosage levels.
[b] Data of Kline et al., Microbiological Associates. q3h = Every 3 h; optimal dose is per injection.
[c] ILS% = Percent increase in median survival time over controls.
[d] The number of long term survivors is given in parenthesis.

Data of three experiments on the influence of the schedule of therapy of adriamycin, taking into account the days of treatment and the number of treatments per day are summarized in Table 3. When adriamycin was administered as a single treatment per day, on a single day, or at two, four or eight day intervals, or on a daily or intermittent daily schedule involving five daily treatments and five days of rest, substantially similar but not outstanding increases in survival time ranging from 40–68% over the untreated controls were observed.

Treatment eight times a day on day 1 (q3h) was considerably more effective than a single treatment on day 1 only. The increase in survival time on the q3h therapy was approximately two to five-fold in two experiments. The q3h schedule for a single day was also more effective than any of the therapeutic schedules in which the drug was administered as single treatments on the day of treatment. When an attempt was made to repeat the q3h therapy at an interval of two, four or eight days, there was no further increase in survival time and in fact an actual reduction of therapeutic efficacy occurred (Table 3, experiments 1, 2). It would appear that the toxicity of adriamycin for the host was limiting and despite an eight day recovery period, a repeated course of q3h therapy was not sufficiently tolerated.

In an attempt to adjust for this an experiment was conducted in which a recovery interval of up to 14 days (q3h days 1,15) was provided between therapies (Table 3, experiment 3). For these schedules on days 1 and 9; 1 and 11; 1 and 13; and 1 and 15, there were no further increases in survival time over that observed with the q3h schedule on day 1 only. The limitation of host toxicity was supported by an additional observation in this experiment that for all of the schedules employed, administration of the optimal dosage resulted in delayed lethality for non-leukemic animals. Although lower dosages were tolerated, the therapeutic effectiveness was not maintained.

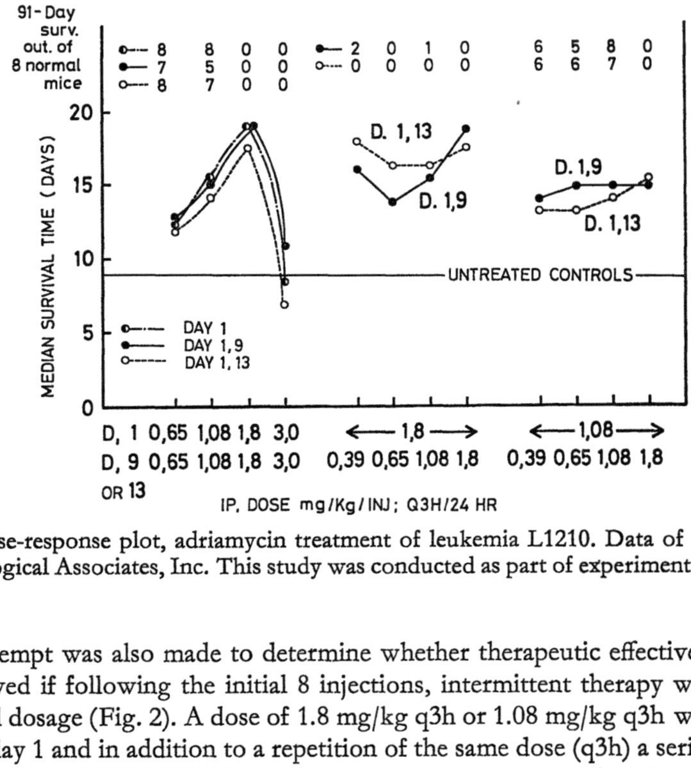

Fig. 2. Dose-response plot, adriamycin treatment of leukemia L1210. Data of Kline et al., Microbiological Associates, Inc. This study was conducted as part of experiment 3 in Table 3

An attempt was also made to determine whether therapeutic effectiveness could be improved if following the initial 8 injections, intermittent therapy were applied at reduced dosage (Fig. 2). A dose of 1.8 mg/kg q3h or 1.08 mg/kg q3h was administered on day 1 and in addition to a repetition of the same dose (q3h) a series of lower doses ranging to 0.39 mg/kg (also q3h) was administered on days 9 or 13. In no instance was there further improvement in therapeutic response as compared with administration every 3 h for a single day. When the second dosage administered on day 9 or 11 was low enough so that host mortality did not occur in non-leukemic mice, there was apparently sufficient loss of therapeutic effectiveness so that no further advantage was obtained in therapy.

Both adriamycin and daunomycin have demonstrated schedule dependency in therapy [6, 7]. For both drugs the outcome of therapy may be influenced by the route of administration of the leukemic cells as well as by the route of injection of the drug. Summary data on "route dependency" for adriamycin and daunomycin are summarized in Tables 4 and 5. When the leukemic cells were inoculated intraperitoneally the intraperitoneal route appeared to be the most effective route for administration of adriamycin. There was little loss of therapeutic effectiveness on subcutaneous administration of adriamycin but a more definite reduction in activity on intravenous administration of the drug. On oral administration adriamycin was ineffective.

Table 4. Effect of drug route on adriamycin and daunomycin activity against leukemia L1210[a] [7]

Drug route	Schedule days	Leukemic cells inoculated intraperitoneally			
		Adriamycin		Daunomycin	
		L1210 OD	ILS%	OD	ILS%
I.P.	1–5, 11–15, 21–25	2.0	63	1.0	31
S.C.	1–5, 11–15, 21–25	2.0	54	4.0	36
Oral	1–5, 11–15, 21–25	8.0	4	8.0	0

[a] BDF_1 male mice (18–26 g). I.P. inoculum 10^5 cells. OD = Optimal dose/mg/kg/injection

Table 5. Effect of drug route on adriamycin and daunomycin activity against leukemia L1210[a]

Drug route	Exp. No.	Schedule days	Leukemic cells inoculated I.P.				Leukemic cells inoculated I.V.			
			Adriamycin		Daunomycin		Adriamycin		Daunomycin	
			OD	ILS%	OD	ILS%	OD	ILS%	OD	ILS%
I.P.	1	1–5 daily	2.0	62	1.0	37	4.0	25	4.0	6
	2	1–5 daily	2.0	39	2.0	22	4.0	6	8	0
I.V.	1	1–5 daily	4.0	37	4.0	25	2.0	19	4.0	25
	2	1–5 daily	4.0	22	4.0	11	4.0	11	8.0	28

[a] CDF_1 male mice (20–26 g) I.P. inoculum 1×10^5 cells, I.V. inoculum 1×10^4 cells. OD = Optimal dose/mg/kg/injection.
Data of Kline et al., Microbiological Associates.

Adriamycin was less effective in the treatment of intravenously inoculated leukemic cells as compared with intraperitoneally inoculated cells. Following intravenous inoculation of the leukemic cells adriamycin appeared only moderately effective on intravenous or intraperitoneal administration in increasing the survival time of the leukemic animals. Thus it is indicated that there is "route dependency" of adriamycin both for the route of leukemic inoculation and route of drug administration.

With daunomycin the "route dependency" for route of leukemic inoculation and route of drug administration appeared similar to that of adriamycin (Tables 4, 5). On intraperitoneal leukemic inoculation, intraperitoneal and subcutaneous drug administration were most effective. For the intraperitoneal inoculation route there apparently was some loss of therapeutic effect on intravenous administration and no activity was observed on oral administration. On intravenous inoculation of L 1210 cells there was moderate activity on intravenous daunomycin administration and loss of activity on intraperitoneal drug administration.

For both adriamycin and daunomycin there was diminished drug toxicity on both subcutaneous and oral administration in non-leukemic animals as compared with the intraperitoneal route. This was, however, as indicated above not reflected in any therapeutic advantage. The loss of activity on oral drug administration at tolerated doses is suggestive of lack of absorption of the drugs from the gastrointestinal tract.

The leukemia P 388 system is a somewhat more sensitive system than leukemia L 1210 and identifies as active most of the compounds which are active in the leukemia L 1210 system [8, 9]. However, the antibiotic mithramycin is inactive in the leukemia L 1210 system and attention was focused on the P 388 system when it was determined that it does identify mithramycin as an active compound. The P 388 system is currently being employed in the screening and evaluation of natural products including fermentation products and plant extracts [9]. The system appears particularly useful for the evaluation of activity in extracts where the active principle is in low concentration, and during the process of purification. The activity ratings of a series of natural products are given in Table 6 for both the L 1210 and P 388 systems [6, 7].

Table 6. Activity ratings of natural products in leukemia,L1210 and leukemia P-388[a] [8, 9]

Name	L1210 ILS%	P−388 ILS%
Adriamycin	63	>445[b]
Daunomycin	58	127
Actinomycin D	45	>175
Mitomycin C	40	150
Vinblastine	40	112
Vincristine	39	>135
DON	25	> 90
Pactamycin	17	45
Mithramycin	13	122

[a] Mouse leukemia P-388 I.P. inoculum 1×10^6 cells. Treatment I.P. every day, days 1−10.
[b] Treatment every 6 hours on day 1 only.

All of the natural products listed were more effective in increasing the survival time of leukemic mice in the P 388 system as compared with the L 1210 system. Adriamycin administered on a favorable schedule every six hours for a single day in the P 388 system was considerably more effective than a series of natural products administered on a daily schedule including mitomycin C, vinblastine and mithramycin. The superiority of adriamycin over daunomycin was illustrated in the P 388 system both on the every 6 h schedule and the every 3 h schedule when the drugs were administered one day following leukemic inoculation (Fig. 3) [7]. Adriamycin was not only more effective in increasing the survival time of the leukemic animals but also yielded a higher incidence of apparently tumor free survivors.

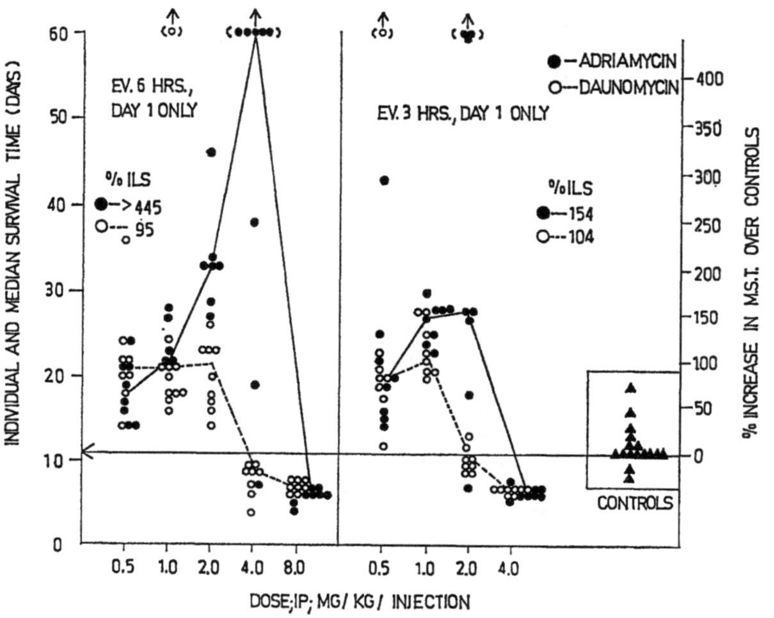

Fig. 3. Mortality distribution plots for the comparative antitumor effect of adriamycin and daunomycin in mice bearing P388 leukemia. Male BDF$_1$ mice (18–26 g) were inoculated (Day 0) ip with 1×10^5 P388 cells from stock DBA/2 mice bearing ascitic P388. Adriamycin or daunomycin was given 1 day after tumor implantation at the doses and schedules indicated. Percent ILS is the percent increase in lifespan over controls. Longterm survivors are indicated by the sign (↑) at the 60 day mark. See Ref. [7]

Because of the limiting toxicity of adriamycin on repetitive administration, even on an intermittent schedule, attention was focused on the possibilities for the employment of other chemotherapeutic agents with adriamycin.

In one experiment a single treatment with adriamycin at 3 days after leukemic inoculation was followed sequentially by a series of treatments with methotrexate from day 4 to death (Table 7). In this study treatment with an optimal dose of adriamycin alone yielded a 60% increase in median survival time while treatment with an optimal dose of methotrexate alone resulted in an 85% increase in median life span. However, when a dose of 14 or 9, or 5 mg/kg of adriamycin was followed by an optimal dose level of methotrexate, the increases in median survival time were 190%, 185% and 155% respectively. The optimal dose of methotrexate in sequential therapy approximated that observed with methotrexate alone.

Table 7. Effect of sequential treatment with adriamycin and methotrexate on the survival time of leukemic (L1210) mice[a]

Adriamycin Day 3 only Optimal dose mg/kg	Methotrexate Day 4 to death Optimal dose mg/kg	Median survival time Days	ILS%
14	—	16	60
—	1.08	18.5	85
Specified dose mg/kg			
14	1.08	29	190
9	1.08	28.5	185
5	1.8	25.5	155
Untreated controls		10	

[a] BDF_1 mice inoculated I.P. with 10^5 L1210 ascites. Treatment I.P. ILS% = percent increase in MST over controls. Data of VADLAMUDI et al., Microbiological Associates.

Table 8. Effect of alternating treatment with adriamycin and methotrexate on the survival time of leukemic (L1210) mice[a]

Adriamycin Day 3—7 13—17 Optimal dose mg/kg	Methotrexate Day 8—12 18—death Optimal dose mg/kg	Median survival time Days	ILS%
1.8	—	15	36
—	1.8	15	36
Specified dose mg/kg			
1.8	0.65	18	64
1.08	1.8	22.5	105
0.65	1.8	16	45
Untreated controls		11	

[a] BDF_1 mice inoculated I.P. with 10^5 L1210 ascites. Treatment I.P. ILS% = Percent increase in MST over controls. Data of VADLAMUDI et al., Microbiological Associates.

In a second experiment involving an alternating form of sequential therapy, adriamycin was administered on days 3—7 following leukemic inoculation (Table 8). This was followed by treatment with methotrexate on days 8—12, by adriamycin on days 13—17 and methotrexate from day 18 death. Here too a therapeutic advantage was observed over adriamycin alone or methotrexate alone. The greatest advantage in therapy was observed with an adriamycin dosage of 1.08 mg/kg alternated with a methotrexate dose of 1.8 mg/kg, where the median survival time was approximately 3 times greater than that observed with adriamycin alone or methotrexate alone.

A similar advantage in sequential therapy has been observed with a combination of adriamycin plus the imidazole carboxamide derivative NSC-45388 [imidazole-4(or 5)-carboxamide, 5(or 4)-(3,3-dimethyl-1-triazeno)-] (DIC) (Table 9). In this experiment adriamycin administered every 3 h on day 1 yielded a 40% increase in median survival time of the leukemic mice and DIC administered daily from day 1—17 gave a 50% increase in median survival time. However, administration of adriamycin on day 1 followed by treatment with DIC from days 3—17 increased the median life span by 125%. It is of interest that with this sequential therapy the optimal dosage for each drug was similar to that observed with the individual drugs, indicating that there was little, if any, additional toxicity for the combination.

Table 9. Effect of sequential treatment with adriamycin and DIC (NSC-45388) on the survival time of leukemic (L1210) mice[a]

Adriamycin Q3H, day 1 Optimal dose mg/kg	DIC Daily, Day 1—17 Optimal dose mg/kg	Median survival time Days	ILS%
1.08	—	14	40[b]
—	180	15	50
	Day 3—17		
1.8	108	22.5	125
Untreated controls		10	

[a] BDF_1 mice inoculated I.P. with 10^5 ascites. Treatment I.P. ILS% = Percent increase in MST over controls. q3h = Every 3 h. DIC (NSC-45388) = imidazole-4 (or 5)-carboxamide, 5 (or 4)-(3,3-dimethyl-1-triazeno-)

[b] The ILS for adriamycin Q3H, day 1 in this experiment was lower than that generally observed. (See Tables 1—3). Data of Kline et al. Microbiological Associates.

In the current study it was determined that the antileukemic effectiveness of adriamycin (and daunomycin) can be influenced by the dosage, route and schedule of administration and by the site of leukemic inoculation. The cumulative toxicity of adriamycin would appear to be limiting and at least one way to avoid this limiting toxicity and thereby achieve a greater therapeutic response is to use adriamycin in combination with other drugs. It may be that adriamycin will find its greatest value in combination chemotherapy. It is suggested that further studies be conducted of factors which may influence and be utilized to improve the chemotherapeutic effectiveness of adriamycin.

The observation that adriamycin which is so closely related structurally to daunomycin is more effective than daunomycin in a number of experimental tumor systems stresses the importance of further analog search for additional effective congeners.

References

1. Di Marco, A., Gaetani, M., Dorigotti, L., Soldati, M., Bellini, O.: Tumori **49**, 203—220 (1963).
2. Arcamone, F., Cassinelli, G., Francheschi, G., Orezzi, P., Penco, S., Solva, A.: La struttura della adriamicina nuovo antibiotico antitumorale (in press).
3. Di Marco, A., Gaetani, M., Scarpinato, B.: Cancer Chemother. Rep. **53**, 33—37 (1969).

4. GOLDIN, A., VENDITTI, J.M., HUMPHREYS, S.R., MANTEL, N.: J. Nat. Cancer Inst. **17**, 203–212 (1956).
5. SELAWRY, O.S., HANANIAN, J., WOLMAN, I.J., ABIR, E., CHEVALIER, L., GOURDEAU, R. DENTON, R., GUSSNÖHOFF, B.D., LEVY, R., BURGERT, O., MILLS, S.D., BLOM, J., JONES, B, PATTERSON, R.B., MCINTYRE, O.R., HAURANI, F.I., MOON, J.H., HOOGSTRATEN, B., KUNG, F.H., SHEEHE, P.R., FREI, E. III, HOLLAND, J.F.: J. Amer. Med. Ass. **194**, 75–81 (1965).
6. VENDITTI, J.M.: A commentary. Treatment schedule dependency of experimentally active antileukemic (L1210) drugs. Cancer Chemother. Rep. 2, 35—59 (1971).
7. SANDBERG, J.S., HOWSDEN, F.L., DI MARCO, A., GOLDIN, A.: Cancer Chemother. Rep. **54**, 1–7 (1970).
8. GOLDIN, A., SERPICK, A.A., MANTEL, N.: Cancer Chemother. Rep. **50**, 173–218 (1966).
9. VENDITTI, J.M., ABBOTT, B.J.: Lloydia **30**, 332–348 (1967).

Antitumor Effects of Adriamycin on Yoshida Rat Sarcoma and L1210 Mouse Leukemia — Cross-resistance and Combination Chemotherapy

Akira Hoshino, Taketoshi Kato, Hiroyuki Amo, and Kazuo Ota

Laboratory of Chemotherapy, Aichi Cancer Center, Research Institute, Nagoya, Japan

With 10 Figures

Adriamycin is a new antitumor antibiotic isolated from cultures of *Streptomyces peucetius var. caesius* [1]. The structural formula of adriamycin differs from daunomycin only in the substitution of a hydroxyl group for a hydrogen atom in the acetyl radical of the aglycone moiety of daunomycin [2]. Adriamycin was reported to possess a higher therapeutic index than daunomycin in serval experimental tumor systems [3, 4]. This paper deals with cross-resistance studies using various resistant sublines of Yoshida sarcoma in rats and L1210 leukemia in mice. Combination chemotherapy of adriamycin with various antitumor agents were also studied in L1210 leukemia.

Materials and Methods

Drug

The drugs used in this study were supplied from following sources: adriamycin, mitomycin C, T-243 (1-hydrox-2-amino-7-methoxy·mitosene·HCl) and 5-fluorouracil: Kyowa Hakko Kogyo Co., Ltd., daunomycin: Meiji Seika Co., Ltd., cytosine arabinoside: Nippon Shinyaku Co., Ltd., methotrexate: courtesy of Dr. D. J. Hutchison of Sloan-Kettering Institute for Cancer Research in New York, USA, endoxan (cytoxan) and vinblastine sulfate: Shionogi & Co., Ltd., 6-mercaptopurine: Morishita Pharm. Co., Ltd., carbazilquinone: Sankyo Co., Ltd., chromomycin A_3 and busulfan: Takeda Chemical Industries Ltd., thio-TEPA: Sumitomo Chemical Co., Ltd., acetylkidamycin: Daiichi Seiyaku Co., Ltd., actinomycin D: Nippon Merk Banyu Co., Ltd.

All drugs were dissolved in saline or suspended in a 0.5% solution of carboxymethylcellulose.

Yoshida Sarcoma

Yoshida ascites sarcoma was taken from Donryu rat and a suspension made in saline. One million cells in 0.2 ml solution were injected intraperitoneally (ip) into each female Donryu rat weighing 100—120 gm. There were 5 rats in each group in this study. Daily ip injections were given from 3 days after inoculation of the sarcoma to a total of 10 doses. The effects of chemotherapy were evaluated by per-cent increase in average survival time (A.S.T.) over saline treated control. Our criteria for the evaluation of chemotherapy have been reported in elsewhere [5].

L1210 Leukemia

L1210 mouse leukemia (L1210/V) [6] was obtained as an ascites from Dr. Dorris J. Hutchison in Sloan-Kettering Institute for Cancer Research, New York, USA and was serially maintained in our laboratory in BDF_1 (C57BL/6 × DBA/2) mice. One million leukemic cells in saline suspension (0.2 ml) were injected ip into BDF_1 mice weighing 18—22 gm. Treatment was started 24 hours after the inoculation with leukemia and continued once every other day for a total of 10 doses. The evaluation of antitumor effects in L1210 leukemia has been reported [5]. There were 5 mice in each group in the most of experiments and 10 mice per group in the two-drug or three-drug combinations.

Drug Resistant Tumors

Drug-resistant sublines of Yoshida sarcoma and L1210 leukemia were selected from the original sarcoma and leukemia cell line by serial transplantation with drug treatment. The resistant sublines are serially maintained in the presence of drugs.

Cross-Restistance and Collateral Sensitivity

Cross-resistance and collateral sensitivity were evaluated by the effects (per cent increase in A.S.T.) of an optimal dose of the drugs against the resistant sublines compared with the effect of the original line [7].

Combination Chemotherapy

Combination chemotherapy with 2 drugs in L1210 leukemia was studied using one-half of the optimal dose of the drugs. The optimal dose (the dose which produces a maximum increase in A.S.T. of treated mice over the control saline-treated mice) of each drug were determined by titration experiments starting from a maximum tolerated dose of the drugs. The optimal doses were 3 mg/kg for amethopterin (MTX), 40 mg/kg for 6-mercaptopurine (6MP), 25 mg/kg for 5-fluorouracil (FU), 2 mg/kg for mitomycin C (MC), 1 mg/kg for daunomycin (DM), 1.5 mg/kg for vinblastine (VLB), 20 mg/kg for cytosine arabinoside (CA), and 0.5 mg/kg for carbazilquinone (CQ) respectively. The percent increase in A.S.T. obtained with optimal doses of drugs ranged between 81% and 131% (Fig. 1).

In the combination experiments the drugs were injected simultaneously, the drugs being mixed just prior to injection.

The effect of the combination therapy was evaluated by comparison with the single use of each drug in optimal dose.

Toxicity

The toxicity of the drugs in chemotherapy experiments was determined by the body weight changes of animals at day 7; when more than 10 gm loss of body weight of rats or more than 2 gm loss of body weight of mice was observed at day 7, the chemotherapy was considered to be toxic.

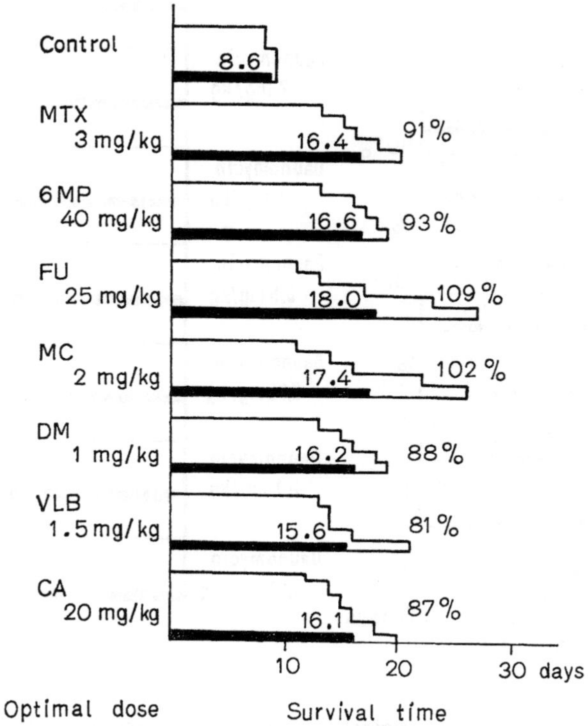

Fig. 1. Original L1210 leukemia. The effect of optimal doses of the various drugs are shown in Fig. 1, as expressed by survival curves. Numbers on the black bar indicate average survival time of leukemia-bearing mice. Percentages indicate the per-cent increase in A.S.T. over the saline treated control.

MTX amethopterin; 6MP 6-mercaptopurine; FU 5-fluorouracil; MC mitomycin C; DM daunomycin; VLB vinblastine sulfate; CA cytosine arabinoside; CQ carbazilquinone

Results

Antitumor Effects and Cross-Resistance in Yoshida Sarcoma

The antitumor effects of adriamycin and daunomycin on the original line of Yoshida sarcoma are shown in Fig. 2 and 3. From the data of Fig. 2 adriamycin showed high therapeutic effects on Yoshida sarcoma. The maximum tolerated dose (MTD) of adiramycin was 4 mg/kg and the minimum effective dose (MED) was less than 0.01 mg/kg. The MTD was 4 mg/kg and MED was 0.05 mg/kg for daunomycin (Fig. 3). The therapeutic index (MTD/MED) of adriamycin was more than 400 times, which was much higher than that of daunomycin (80 times). The optimal dose of adriamycin for Yoshida sarcoma was 1 mg/kg in our treatment schedule.

Cross-resistance of adriamycin in the 10 sublines of Yoshida sarcoma resistant to various drugs was studied and the data are summarized in Table 1. The induction

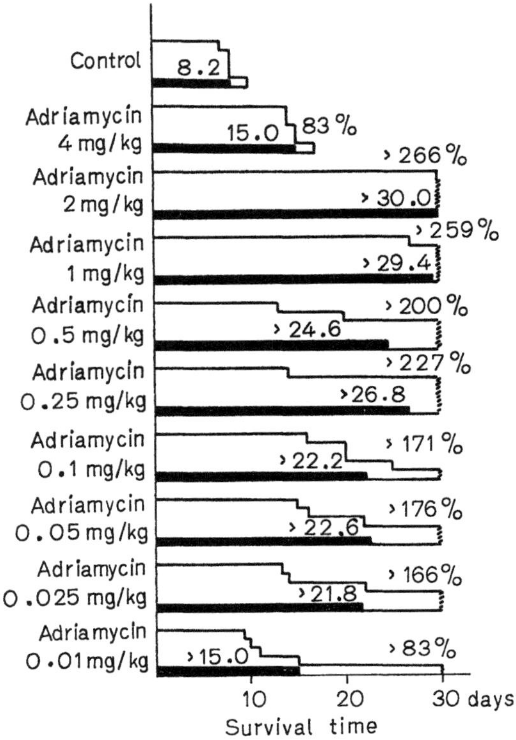

Fig. 2

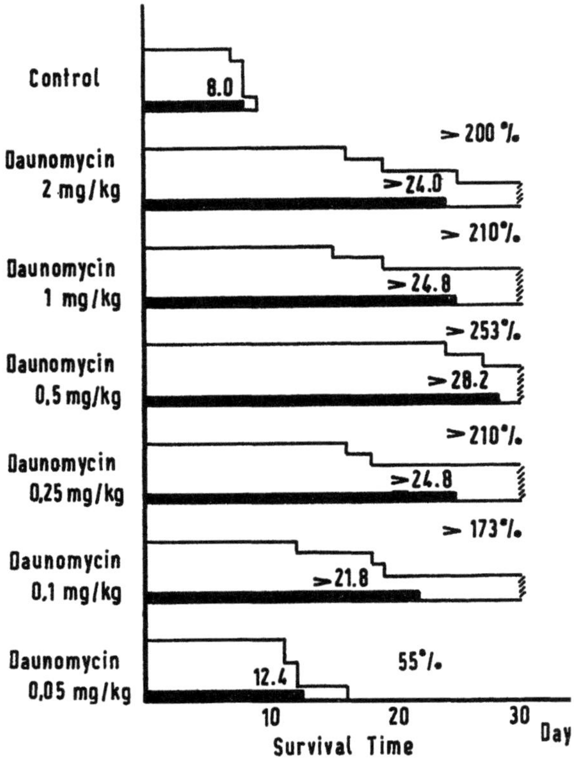

Fig. 3

Fig. 2. Effect of adriamycin on original line of Yoshida sarcoma. Fig. 2 shows survival curves of Yoshida sarcoma-bearing Donryu rats treated by adriamycin. The numbers on the black bar indicate average survival time (A.S.T.) of rats. The percentages express percent increase in A.S.T. over the saline-treated controls. The survival time of rats which survived more than 30 days is calculated as 30 days and indicated as numbers with (>)

Fig. 3. Effect of daunomycin on original line of Yoshida sarcoma. Fig. 3 shows survival curves of Yoshida sarcoma rats treated by daunomycin

of resistance to daunomycin and chromomycin A_3 was fairly difficult and slow compared with other drugs. Resistance to daunomycin was established after 40 transfer generations. Resistance to endoxan, busulfan and thio-TEPA developed at about 12–13 transfer generations. Resistance to 6-mercaptopurine and 5-fluorouracil was completed at about 6—7 generations. On the other hand, resistance to mitomycin C was established by only one transfer generation [8, 9].

At doses of 1.0 mg/kg and 0.5 mg/kg, adriamycin was effective in all resistent sublines except the subline resistant to daunomycin (Table 1), i.e. the daunomycin-resistant subline showed cross-resistance to adriamycin. Increased doses of adriamycin could not overcome the cross-resistance of the daunomycin subline.

Cross-resistance experiments were carried out among several antitumor antibiotics and the results are shown in Table 2.

Table 1. Effect of adriamycin on original and resistant lines of Yoshida sarcoma

	Adriamycin 1.0 mg/kg	Adriamycin 0.5 mg/kg
Original Yoshida sarcoma	++	++
Mitomycin C 0.5 mg/kg Resistant line[a]	++	++
T-243 2 mg/kg Resistant line	++	++
Endoxan 5 mg/kg Resistant line	++	++
Busulfan 10 mg/kg Resistant line	++	++
Thio-TEPA 1 mg/kg Resistant line	++	++
Chrommycin A_3 0.01 mg/kg resistant line	++	++
Daunomycin 0.5 mg/kg resistant line	−	−
6 Mercaptopurine 10 mg/kg resistante line	++	++
5 Fluorouracil 10 mg/kg resistant line	++	++

− < 50%. ± 50−100%. + > 100%. ++ > 100% with 30 day survivor. % increase in average survival time.

Cross-resistance of adriamycin in various resistant sublines of Yoshida sarcoma was studied at the dose of 0.1 mg/kg (optimal dose) and 0.5 mg/kg of adriamycin.

[a] The mitomycin C-resistant subline was established by subcutaneous injection of 0.5 mg/kg of mitomycin C.

Table 2. Effects of various anti-tumor antibiotics on original and resistant lines of Yoshida sarcoma

	SS	MC-R	DM-R	TM-R
Daunomycin	++	++	−	++
Adriamycin	++	++	−	++
Chromomycin A_3	++	++	−	−
Acetyl-Kidamycin	+	−	−	+
Actinomycin D	++	++	−	±

++ Increase in A.S.T. > 100% with 30 day survivor. + Increase in A.S.T. > 100%. ± Increase in A.S.T. 100%−50%. − Increase in A.S.T. < 50%.

Yoshida sarcoma lines
SS Original line of Yoshida sarcoma.
MC-R Mitomycin C 0.5 mg/kg resistant subline.
DM-R Daunomycin 0.5 mg/kg resistant subline.
TM-R Chromomycin A_3 0.5 mg/kg resistant subline.
Dose
Daunomycin 0.5 mg/kg
Chromomycin A_3 0.025 mg/kg,
Actinomycin D 0.05 mg/kg,
Adriamycin 0.5 mg/kg,
Acetyl-kidamycin 1.0 mg/kg.

The cross-resistance pattern of adriamycin was similar to that of daunomycin. The daunomycin-resistant subline also showed cross-resistance to chromomycin A_3, acetyl-kidamycin and actinomycin D. The cross-resistance pattern of chromomycin

A_3 resembled to actinomycin D. The cross-resistance pattern for acetyl-kidamycin is between that of the group of adriamycin and daunomycin and the group of chromomycin A_3 and actinomycin D. From these results, it might be assumed that the modes of action of adriamycin and daunomycin are very similar and are somewhat different from acetyl-kidamycin, chromomycin A_3 and actinomycin D, which all show strong binding activity to DNA of tumor cells [10, 11].

Antitumor Effects and Cross-Resistance in L1210 Leukemia

The antitumor effect of adriamycin on the original line of L1210 mouse leukemia is shown in Fig. 4. The maximum tolerated dose (MTD) of adriamycin was 4 mg/kg and the minimum effective dose (MED) was 0.25 mg/kg in our experimental schedule.

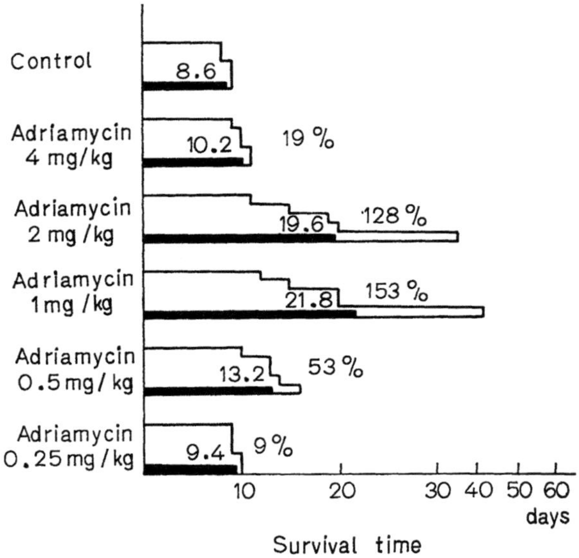

Fig. 4. Effect of adriamycin on L1210 leukemia. Fig. 4 shows survival curves of L1210 leukemia-bearing BDF_1 mice. The scale for survival time on the abscissa is reduced by 1/3 after 30 days

The therapeutic index of adriamycin was 16 times and was twice as high as that of daunomycin (× 8 [5]). The percent increase in A.S.T. by the treatment with the optimal dose (1 mg/kg) of adriamycin was 153%, much higher than that (88%) of daunomycin (see Fig. 1). The cross-resistance studies on adriamycin were carried out in four resistant sublines of L1210 leukemia. The results are summarized in Table 3. Complete cross-resistance to adriamycin was observed in the subline resistant to daunomycin. Partial or relative cross-resistance to adriamycin and daunomycin was seen in the subline resistant to 6-mercaptopurine. The amethopterin-resistant subline showed collateral sensitivity to both adriamycin and daunomycin. The effects of both drugs on the 5-fluorouracil-resistante subline are similar to original L1210 leukemia. The relative cross-resistance to adriamycin and daunomycin in 6-mercaptopurine-resistant subline might be explained by a decrease of permeability to the drugs or decrease in binding capacity on the cell surface of the resistant leukemia cells. The presence of collateral sensitivity to adriamycin and daunomycin in the amethopterin-resistant subline is of interest and might be an advantage for sequential administration of the drugs in clinical combination chemotherapy.

Table 3. Comparative effect of adriamycin and daunomycin on resistant lines of L1210 leukemia

	Original L1210 leukemia	Amethopterin resistant 3 mg/kg	6 Mercaptopurine resistant 40 mg/kg	5 Fluorouracil resistant 25 mg/kg	Daunomycin resistant 1 mg/kg
Adriamycin 2 mg/kg	100	230	50	110	0
Adriamycin 1 mg/kg	100	240	50	70	0
Daunomycin 2 mg/kg	100	320	60	130	0
Daunomycin 1 mg/kg	100	90	50	100	0

0∼ 20 Cross resistance
30∼ 70 Relative cross resistance
80∼120 Sensitive
130∼ Collateral sensitivity.

Table 3 shows comparative antitumor activity of adriamycin and daunomycin in L1210 leukemia.

The numbers were calculated from the percent increase in A.S.T. of treated mice bearing resistant leukemia as 100 for the percent increase in A.S.T. of original L1210 leukemia.

Numbers below 70 indicate the presence of cross-resistance and numbers above 130 indicate collateral sensitivity of resistant leukemia to the drugs.

Combination Chemotherapy with Two Drugs in L1210 Leukemia

The experiments of combination chemotherapy with 2 drugs were carried out using one-half of the optimal dose of the drugs, i.e. 1.5 mg/kg for amethopterin (MTX), 20 mg/kg for 6-mercaptopurine (6MP), 12.5 mg/kg for 5-fluorouracil (FU), 1 mg/kg for mitomycin C (MC), 0.5 mg/kg for daunomycin (DM), 0.75 mg/kg for vinblastine (VLB), 10 mg/kg for cytosine arabinoside (CA), 0.25 mg/kg for carbazilquinone (CQ) and 0.5 mg/kg for adriamycin (AM), respectively (Fig. 5). The increase in A.S.T. was 83% by the combination of AM + MTX, 100% by AM + 6MP, 113% by AM + FU, more than 225% with a 60 day survivor by AM + MC, 52% by AM + VLB, more than 264% with a 60 day survivor by AM + CA, and 165% by AM + CQ respectively (Fig. 5).

Thus, the combination AM + MC and AM + CA, provided remarkable tumor inhibitory effects. These effects of the combinations were considered to be more than additive or synergistic antitumor activity when they were compared to the effect of the optimal dose of each of the drugs alone (Fig. 1). The combinations

AM + MTX, AM + 6MP, AM + FU and AM + CQ were considered to show an additive effect of the two drugs. The combination AM + VLB was less than the additive effect of the two drugs.

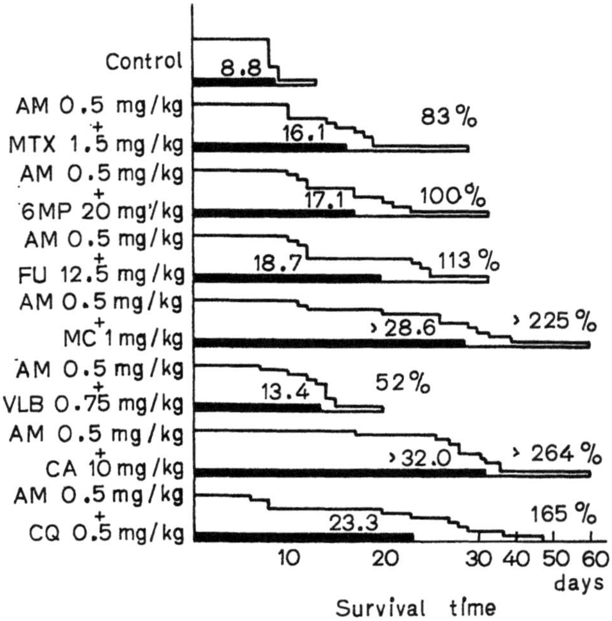

Fig. 5. Effect of combination chemotherapy of adriamycin with various agents on L 1210 leukemia. AM adriamycin; MTX amethopterin; 6MP 6-mercaptopurine; FU 5-fluorouracil; MC mitomycin C; VLB vinblastine sulfate; CA cystoine arabinoside; CQ carbazilquinone. The drugs were given simultaneously i.p. to BDF_1 mice. There were 10 mice in each group

The effects of two-drug combination therapy are summarized in Table 4. Of 35 combinations with two drugs, the combinations MC + FU, FU + CA, CA + MC, DM + CA, 6MP + CA, CQ + CA, AM + CA and AM + MC produced therapeutic synergism. Most of the other combinations showed an additive effect (Table 4).

Combination Chemotherapy with Three Drugs

Of the two-drug combinations, MC + FU, FU + CA and CA + MC showed a synergistic effect. The drugs (MC, FU and CA) seem to be in a triangular interrelation, so that a triple combination of these three drugs was carried out using an equilateral triangle design of dose schedule (Fig. 6 [12, 13]). When the optimal doses of the three drugs are placed on the vertices of the triangle, the height of the vertex (length of perpendiculars) expresses the optimal dose. For any point within the triangle, the sum of the heights of three perpendiculars corresponds to the optimal dose of each of the drug alone. This constancy obtains not only for points within the triangle, but also for the points along the sides (equivalent to two drug combinations) and on the vertices of the triangle (single-drug treatments). The results (per-cent increase in A.S.T.) are expressed three dimensionally, by the length of the bars which stand on the triangle to give a stereo view (Fig. 7) [13, 14].

Table 4. Effect of combination chemotherapy on L1210 leukemia (2 agents)

	6MP	FU	MC	DM	VLB	CA	CQ	AM
MTX	±	+	+	±	±	+	+	+
6MP		+	+	±	±	⧺	+	+
FU			⧺	+	+	⧺	+	+
MC				+	+	⧺	+	⧺
DM					±	⧺	+	
VLB						+	±	±
CA							⧺	⧺
CQ								+

⧺ Synergism. + Additive effect. ± Less than additive.

When the percent increase in A.S.T. of combination therapy was considerably greater than the percent increase in A.S.T. of the optimal dose of each of the drugs alone, the combination was considered to be synergistic.

MTX amethopterin; 6MP 6-mercaptopurine; FU 5-fluorouracil; MC mitomycin C; DM daunomycin; VLB vinblastine; CA cytosine arabinoside; CQ carbazilquinone; AM adriamycin.

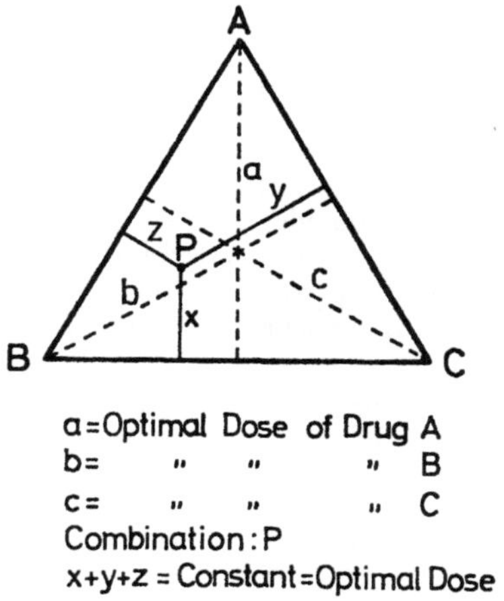

Fig. 6. Dose schedule for combination chemotherapy of three drugs. Goldin et al.: Cancer Res., **28**, 950 (1968)

The maximum therapeutic effect, a more than 522% increase in A.S.T. with four out of five mice as 60 day survivors, was observed with the combination of one-third of the optimal dose of each of three drugs, which corresponds to the point at the center of the triangle. However, the increse in A.S.T. by the treatment of MC, FU and CA alone was 114%, 124% and 94% respectively. Combination of one-fourth of MC plus one-fourth of FU plus one-half of CA and one-half of MC plus one-fourth of FU plus one-fourth of CA also showed high therapeutic synergism.

Toxic effects (more than 2 gm weight loss of mice at day 7) were not proportionally increased for any effective combinations of the three drugs.

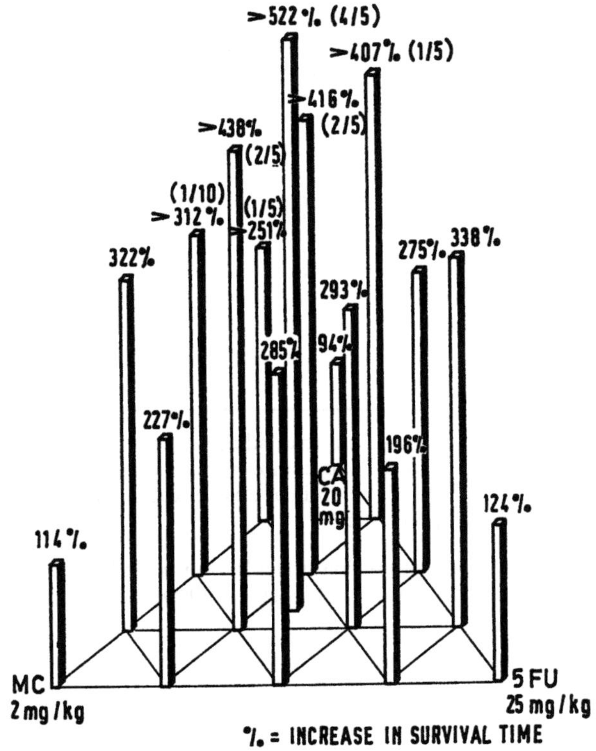

Fig. 7. Effect of combination chemotherapy of mitomycin C, 5-fluorouracil and cytosine arabinoside on L1210 leukemia. ( ) number of 60 day survivors in the group of treated mice. Per-cent increase in A.S.T. of the group with 60-day survivors was calculated including the survivors as they survived for 60 days.

> indicates the presence of 60-day survivors.

MC mitomycin C; CA cytosine arabinoside; 5FU 5-fluorouracil

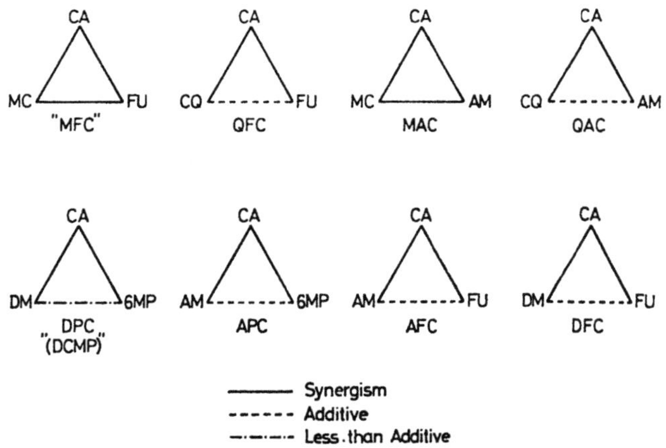

Fig. 8. Promising combination chemotherapy of three drugs in L1210 leukemia.
MC mitomycin C; CA cytosine arabinoside; FU 5-fluorouracil; 6MP 6-mercaptopurine; CQ carbazilquinone; AM adriamycin; DM daunomycin.
M mitomycin C; Q carabazilquinone; F 5-fluorouracil; A adriamycin; C cytosine arabinoside; D daunomycin; P 6-mercaptopurine. MFC mitomycin C + 5-fluorouracil + cytosine arabinoside; DCMP daunomycin + cytosine arabinoside + 6-mercaptopurine + prednisolone

Fig. 8 shows how schemes of promising combination chemotherapy for three drugs can be deduced from the results of two-drug combinations. A solid line expresses the presence of a synergistic effect between two drugs and a dotted line indicates an additive effect of two drugs. In this scheme, the combinations MC + FU + CA "MFC" and DM + 6MP + CA "DCMP" were already under going clinical trials and these combinations (MFC and DCMP) were proved to posses high therapeutic effects on human malignancies [15, 16]. Since the two-drug combinations CA + DM and CA + 6MP, or CA + CQ and CA + AM showed therapeutic synergism (Fig. 8, Table 4), triple combinations of DM + CA + 6MP (Fig. 9) [17] and CQ + CA + AM (Fig. 10) were carried out in L1210 leukemia.

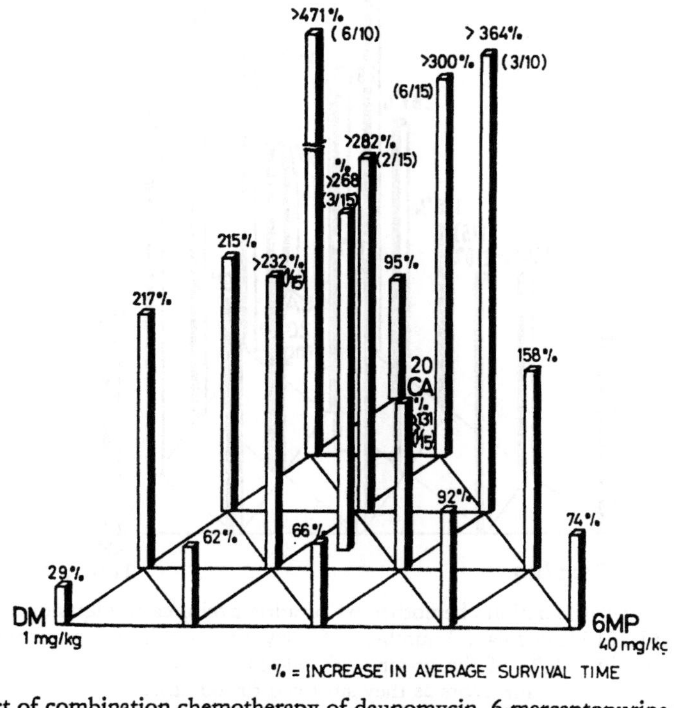

Fig. 9. Effect of combination chemotherapy of daunomycin, 6-mercaptopurine and cytosine arabinoside on L1210 leukemia. ( ) number of 60-day survivors in the group of treated mice. Per-cent increase in A.S.T. of the group with 60-day survivors was calculated including the survivors as they survived for 60 days.
> indicates the presence of 60-day survivors.
DM daunomycin; CA cytosine arabinoside; 6MP 6-mercaptopurine

The combinations of these three drugs showed high therapeutic synergism. In Fig. 9, the triple combination of one-third of the optimal dose of DM, CA and 6MP produced a more than 268% of increase in A.S.T. with three 60-day survivors out of 15 mice. The combinations of one-half of DM plus one-fourth of CA plus one-fourth of 6MP, and of one-fourth of DM plus one-half of CA plus one-fourth of 6MP, showed high therapeutic synergism. However, by the proportion of the drugs, a two drug combination CA + DM or CA + 6MP might be better than the three-drug combination. The combinations of three-fourths of CA plus one-fourth of DM showed high therapeutic synergism, i.e. a more than 471% of increase in A.S.T. with six 60-day survivors. The combinations of three-fourths of CA plus one-fourth of 6MP and one-half of CA plus one-half of 6MP were highly synergistic. These results of the two-drug combinations can be seen from the bars on the sides of the triangle.

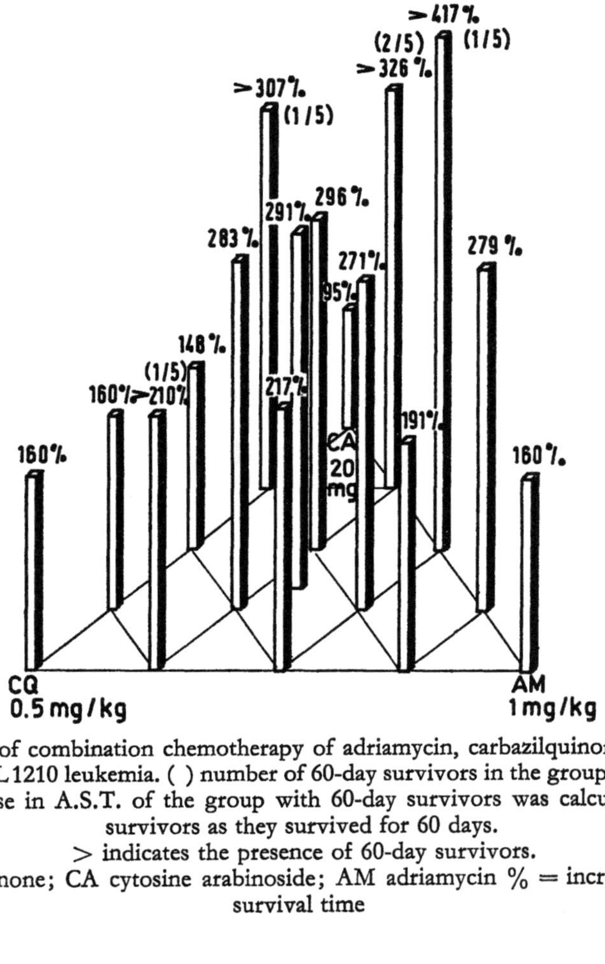

Fig. 10. Effect of combination chemotherapy of adriamycin, carbazilquinone and cytosine arabinoside on L1210 leukemia. ( ) number of 60-day survivors in the group of treated mice. Per-cent increase in A.S.T. of the group with 60-day survivors was calculated including survivors as they survived for 60 days.
> indicates the presence of 60-day survivors.
CQ carbazilquinone; CA cytosine arabinoside; AM adriamycin % = increase in average survival time

The triple combination CQ + CA + AM is shown in Fig. 10. The percent increase in A.S.T. was 291% for the combination of one-third of CQ plus one-third of CA plus one-third of AM, 283% for one-half of CQ plus one-fourth of CA plus one-fourth of AM, 269%, for one-fourth of CQ plus one-half of CA plus one-fourth of AM, and 271% for one-fourth of CQ plus one-fourth of CA plus one-half of AM, respectively. However, the two-drug combinations of three-fourths of CA plus one-fourth of CQ, three-fourths of CA plus one-fourth of AM, and one-half of CA plus one-half of AM, also showed a high therapeutic effect of more than 307% with one out of 5 mice as 60-day survivors for the first combination, more than 326% with two 60-day survivors for the second combination, and more than 417% with one 60-day survivors for the third combination.

Since these three drugs are known to be effective against human leukemia and solid tumors, the combination of CQ + CA + AM might be a promising therapy for neoplasia in man.

Discussion

In these studies, it was clearly demonstrated that adriamycin was more effective than daunomycin in the therapeutic indicies of rats bearing Yoshida sarcoma and of mice bearing L1210 leukemia. The therapeutic index (MTD/MED) for adriamycin in Yoshida sarcoma was more than 400times which was much higher than that (80 times) for daunomycin. The therapeutic index for adriamycin in L1210 leukemia was 16 times and for daunomycin was 8 times.

The higher therapeutic index of adriamycin compares with daunomycin was confirmed in the series of experimental tumor system including S180, Ehrlich ascites carcinoma, Oberling-Guérin-Guérin carcinoma, a nitroso-methylurea-induced lymphosarcoma of C3Hf mice [3], L1210 leukemia and P388 lymphocytic leukemia [3, 4, 18].

Cross resistance of adriamycin was observed in the daunomycin resistant sublines of Yoshida sarcoma and of L1210 leukemia [11]. Cross-resistance patterns of adriamycin, daunomycin, chromomycin A_3, actinomycin D and acetyl-kidamycin were of interest for differences in each of the drugs. From these patterns, the antibiotics are classified in three groups which are the group of adriamycin and daunomycin, the group of chromomycin A_3 and actinomycin D, and the group of acetyl-kidamycin.

Since daunomycin binds to deoxyribonucleic acid [19, 20] and inhibits DNA-dependent RNA synthesis [21, 22], and also DNA-dependent DNA synthesis [23, 24], the modes of action of adriamycin and daunomycin are fairly close. However, the bindings to DNA in actinomycin D [25], chromomycin A^3 [26] and acetyl-kidamycin [10] are somewhat different from those of daunomycin and adriamycin.

In the combinations of two drugs, adriamycin plus mitomycin C and adriamycin plus cytosine arabinoside provided remarkable therapeutic synergism. Of 35 combinations with two drugs, eight combinations were synergistic [14]. From these data, combination therapy of three drugs including daunomycin and adriamycin looks promising.

The triple combinations with carbazilquinone, cytosine arabinoside and adriamycin showed therapeutic synergism.

The excellent synergistic effect of the combination of these three drugs, demonstrated in this paper, might be attributed different biochemical sites of action of these drugs, all of which inhibit DNA synthesis of tumor cells [24, 27, 28].

Since the triple combination of mitomycin C, 5-fluorouracil and cytosine arabinoside, or the quadruple combination of daunomycin, cytosine arabinoside, 6-mercaptopurine and prednisolone, whose strong tumor-inhibiting effects were confirmed in L1210 leukemia, were highly effective against human gastrointestinal carcinoma (former combination) [15] and against acute granulocytic leukemia in adults (later combination) [16], we conclude that the triple combination of carbazilquinone, cytosine arabinoside and adriamycin deserves a clinical trial.

Acknowledgments

The authors would like to express our sincere gratitude to Dr. D. J. Hutchison, SKI, New York, Dr. A. Goldin and Mr. S. M. Poiley, NCI, Bethesda for their fruitful discussions and for kindly supplying tumors and mice. The authors are indebted to Mrs. M. Ishikawa, Mrs. Y. Terashima, Miss. K. Nishida and Miss. T. Morita for their skilful assistance and to Dr. K. Kimura, NCC, Tokyo for continued interest and encouragement.

This work has been supported by research grants from the Ministry of Public Health and Welfare and the Ministry of Education, Japan.

References

1. Di Marco, A., Gaetani, M., Dorigotti, L., Soldati, M., Bellini, O.: Tumori **49**, 203–220 (1963).
2. Arcamone, F., Cassinelli, G., Franceschi, G., Prezzi, P., Penco, S., Solva, A.: La struttura della adriamycin nuovo antibiotico antitumorale (in press).
3. Di Marco, A., Gaetani, M., Scarpinato, B.: Cancer Chemother. Rep. **53**, 33–37 (1967).
4. Sandberg, J. S., Hawsden, F. L., Di Marco, A., Goldin, A.: Cancer Chemother. Rep. **54**, 1–7 (1970).
5. Hoshino, A., Kato, T., Amo, H., Ota, K.: Studies on resistance to anticancer agents (VII), Antitumor effect of daunomycin on Yoshida sarcoma and L1210 leukemia. Proc. Japanese Cancer Association. The 27th Annual Meeting, October, p. 226, 1968, (Tokyo).
6. Hutchison, D. J., Robinson, D. L., Martin, D., Ittenson, O. L., Dillernberg, J.: Cancer Res. **22**, 57–72 (1962).
7 Hoshino, A.: J. Jap. Soc. Cancer Ther. **4** (Suppl), 28–34 (1969).
8. Hoshino, A.: Experimental studies on the resistance to mitomycin C. 5th International Congress of Chemotherapy. Proceedings, Spitzy, K. H., Haschek, H. (Eds.). Vol. III, p. 637–639. Wien: Verlag der Wiener Medizinischen Akademie 1967.
9. Hoshino, A.: Nippon Rinsho, **27** (Suppl), 108–115 (1969).
10. Hoshino, A., Ishikawa, M., Amo, H., Kato, T., Ota, K.: Studies on resistance to anticancer agents. 10. Biochemical studies on antitumor effect of acetyl-kidamycin. Proc. Japanese Cancer Association. The 29th Annual Meeting, October, p. 231, 1970 (Osaka).
11. Hoshino, A., Kato, T., Amo, H., Ota., K: Studies on resistance to anticancer agents. 13. Antitumor effects of adriamycin on Yoshida sarcoma and L1210 leukemia with special reference to cross resistance and combination chemotherapy. Proc. Japanese Cancer Association, the 30th Annual Meeting, October, p. 117, 1971 (Tokyo).
12. Goldin, A., Venditti, J. M., Mantel, N., Kline, I., Gang, M.: Cancer Res. **28**, 950–960 (1968).
13. Hoshino, A., Kato, T., Amo, H., Ota, K.: Experimental studies on the triple combination therapy (MFC therapy) of mitomycin C, 5-fluorouracil and cytosine arabinoside. Chemotherapy **18**, 384–385 (1970).
14. Hoshino, A., Kato, T., Amo, H., Ota, K.: L1210 mouse lekuemia as an experimental model system for clinical combination chemotherapy. Proc. VIIth International Congress of Chemotherapy, München. Urban & Schwarzenberg in cooperation with Praha. Avicenum (in press).
15. Ota, K., Kurita, S., Nishimura, M., Ogawa, M., Kamei, Y., Imai, K., Ariyoshi, Y., Kataoka, K., Murakami, M., Oyama, A., Hoshino, A., Amo, H., Kato, T.: J. Jap. Cancer Ther. **6**, 267–276 (1971).
16. Kamei, Y., Kurita, S., Ogawa, M., Imai, K., Murakami, M., Ota, K.: Acta Heam. Jap. **33**, 784–785 (1970).
17. Hoshino, A., Kato, T., Amo, H., Ota, K.: Acta Heam. Jap. **33**, 781–782, (1970).
18. Goldin, A.: Some factors influencing the chemotherapeutic effectiveness of adriamycin. International Symposium on Adriamycin, Milan, Italy, 1971.
19. Goldberg, I. H.: Amer. J. Med. **39**, 722–752 (1965).
20. Wand, D. C., Reich, E., Goldberg, I. H.: Science **149**, 1259–1263 (1965).

21. Di Marco, A.: Path, Biol. **15**, 897–902 (1967).
22. Pittillo, R. F., Hunt, D. E.: Cancer Chemother. Rep. **52**, 707–713 (1968).
23. Kessel, D., Botterill, V., Wodinsky, I.: Cancer Res. **28**, 938–941 (1968).
24. Di Marco, A.: Adriamycin: the activity on experimental tumors. International Symposium on Adriamycin. Milan, Sept. 1971.
25. Raven, H. M., Kersten, H., Kersten, W.: Z. Physiol. Chem. **321**, 139–147 (1960).
26. Kersten, W., Kersten, H.: Biochem. Z. **341**, 174–183 (1965).
27. Chu, M. Y., Fischer, G. A.: Biochem. Pharmacol. **11**, 423–430 (1962).
28. Hoshino, A.: On the mechanisms of antileukemic agents studied by means of drug resistance and on the experimental studies of combination chemotherapy. Gann-no-Rinsho, (in press).

Cross-resistance between Adriamycin, Daunomycin and Vincristine in Ehrlich Ascites Tumor *in vivo*

KELD DANØ

Department of Internal Medicine, Finsen Institute, Copenhagen, Denmark

With 6 Figures

Development of resistance is a major problem in cancer chemotherapy, and is difficult to investigate clinically. Most work in this field has therefore been done using experimental tumors as model systems. In the present investigation, the development of resistance to several drugs, including daunomycin and adriamycin, was studied in Ehrlich ascites tumor *in vivo* in mice.

The toxicity of daunomycin and adriamycin was examined in the mice when the drugs were administered by the same route and schedule as later used in the therapy experiments. Fig. 1 shows a log probit plot of the results. Drugs were given ip daily for 5 days, and the number of mice dying within 30 days was recorded. LD 10 was calculated for daunomycin as 1.3 mg/kg (5% confidence limits 1.1-1.6 mg/kg), and for

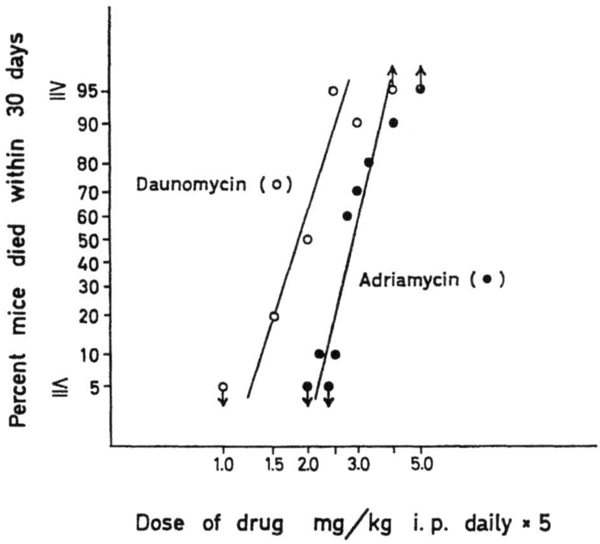

Fig. 1. Log probit plot of toxicity of adriamycin and daunomycin in female mice. 10 mice were tested for each dose, except for daunomycin at 2.5 mg/kg where 20 mice were used

adriamycin as 2.3 mg/kg (5% confidence limits 2.2-2.4 mg/kg). The difference in toxicity was highly significant, adriamycin thus being less toxic than daunomycin, when given by this route and with this schedule. This is the opposite of what has been reported for acute intravenous toxicity in mice, which was found to be higher for adriamycin than for daunomycin [DiMarco et al.: Cancer Chemother. Rep. **53**, 33-37 (1969)]. This suggests a difference in the pharmacodynamic behavior of the two drugs, e. g. in the uptake from the peritoneal cavity or in the rate or site of drug metabolism.

The tumor-inhibiting effect of the two drugs on the wild tumor was studied in therapy experiments, and the results are shown in Fig. 2. Mice were inoculated ip with 15 million tumor cells and divided into groups of 8 – 18, wich were treated intraperitoneally daily for 5 days, starting 24 h after inoculation, with various drug doses up to LD 10. On day 7, the ascites tumor cell volume in each mouse was determined, and the average for each group in percent of control ± SE is indicated in Fig.2.

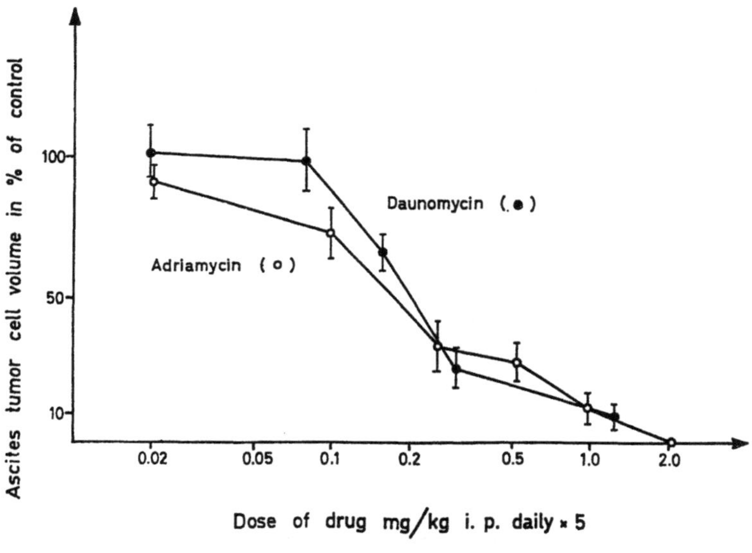

Fig. 2. Response to treatment with adriamycin and daunomycin of a wild Ehrlich ascites tumor. Each dot represents mean of total ascites tumor cell volume 7 days after inoculation of 15×10^6 cells in 8–12 treated mice in percent of mean total ascites tumor cell volume in 18–19 control mice. Vectors indicate SE of the mean values

Both drugs inhibit tumor growth in doses below LD 10, and the effect appears to be equal for the same amounts of the two drugs by weight. Table 1 shows the results of an experiment where the effect of equal doses of the two drugs was compared directly. It is seen that there is no significant difference in effect. It has been reported by DiMarco and collaborators in the study mentioned, that adriamycin was more effective than daunomycin in Ehrlich ascites tumor, when the effect on survival was measured. The present data are not inconsistent with this, but show that in the present system the main difference is that adriamycin is less toxic than daunomycin, while the tumor inhibiting effect is similar.

Table 1. Growth inhibition of wild Ehrlich ascites tumor with daunomycin and adriamycin. For experimental conditions see text to Fig. 2

Drug	Dose (mg/kg, ip daily ×5)	Total ascites tumor cell volume in traeted mice (percent of control ± SE)
Daunomycin	0.5	34 ± 6
Adriamycin	0.5	35 ± 5
Daunomycin	1.0	24 ± 9
Adriamycin	1.0	23 ± 5

As described previously [Danø: Cancer Chemother. Rep. **55**, 133-141 (1971)], resistance to daunomycin was developed by treatment with the drug in 16 weekly passages, the resistant tumor being insensitive to LD 10 of the drug. The resistance was gradually reversible when the treatment was discontinued. Fig. 3 shows the results of therapy with adriamycin on the daunomycin-resistant tumor compared to the wild tumor, and it is seen that the tumor shows cross-resistance to adriamycin.

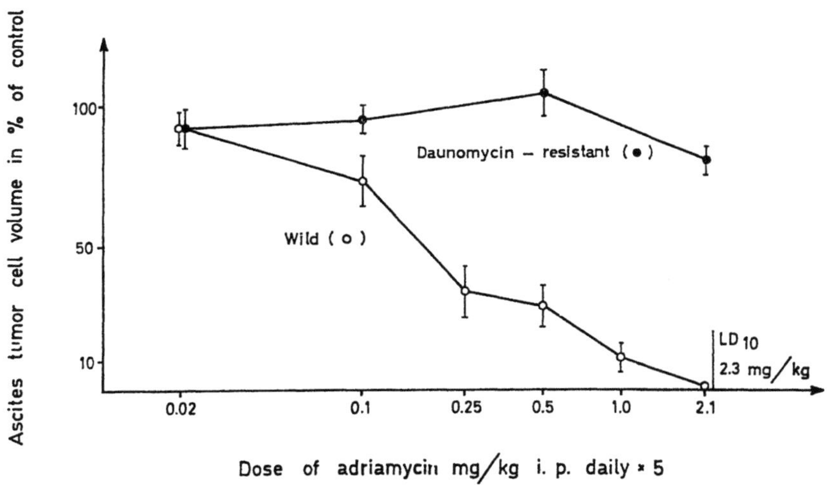

Fig. 3. Response to treatment with adriamycin of a wild Ehrlich ascites tumor and of a subline selected for resistance to daunomycin. For experimental conditions and explantion of signs see text to Fig. 2. Response of the daunomycin-resistant tumor to daunomycin at 1.25 mg/kg ip daily × 5 in the same experiment was inhibition to 98% ± 5% (mean value ± SE) of control

A strain of the wild tumor was treated with adriamycin in 25 weekly passages and was then found to be completely resistant to the drug in doses up to LD 10. A subline of the adriamycin-resistant tumor, which has now been maintained for 25 weeks without adriamycin treatment, shows no loss of resistance, but it is too early to evaluate whether the resistance is stable, as a daunomycin-resistant tumor line in which daunomycin was continued for 46 weeks and then was discontinued showed a gradual loss of resistance, which first started after 30 weeks without daunomycin.

Fig. 4 shows the results of a therapy experiment with daunomycin on the adriamycin-resistant tumor. It is seen that the tumor shows cross-resistance to daunomycin.

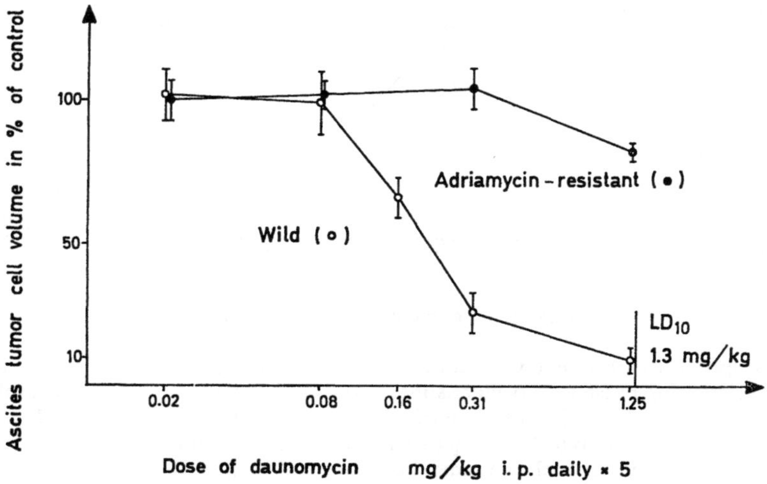

Fig. 4. Response to treatment with daunomycin of a wild Ehrlich ascites tumor and of a subline selected for resistance to adriamycin. For experimental conditions and explanation of signs see text to Fig. 2. Response of the adriamycin-resistant tumor to adriamycin at 2.1 mg/kg ip daily × 5 in the same experiment was inhibition to 88% ± 6% (mean value ± SE) of control

More surprisingly, the adriamycin-resistant tumor also showed cross-resistance to vincristine. There was no change in sensitivity to methotrexate, and – unchanged – no sensitivity to 6-mercatopurine. Collateral sensitivity was found to cytosine arabinoside and BCNU (Danø: Cancer. Chemother. Rep., in press). Relative to the wild tumor the response of the daunomycin-resistant tumor to the drugs mentioned was identical to the findings for the adriamycin-resistant tumor.

Recently, we developed resistance to vincristine by treatment with the drug in a line of our Ehrlich ascites tumor. Fig. 5 shows the response of this tumor to adriamycin, and it is seen that the tumor is cross-resistant to adriamycin. Similarly we have found that the vincristine-resistant tumor is cross-resistant to daunomycin.

The mutual cross-resistance between daunomycin, adriamycin and vincristine is of immediate interest. Cross-resistance to vincristine and adriamycin has also been described in Chinese hamster cells in vitro, in which resistance was developed to daunomycin [Reihm and Biedler: Cancer. Res **31**. 409-412 (1971)], but it remains to be investigated whether cross-resistance between the three drugs is a general phenomenon. An analysis of clinical data for possible cross-resistance between the drugs would be of considerable interest.

The mechanism of resistance has been investigated in the daunomycin-resistant tumor, where a markedly decreased uptake of daunomycin in the resistant cells was

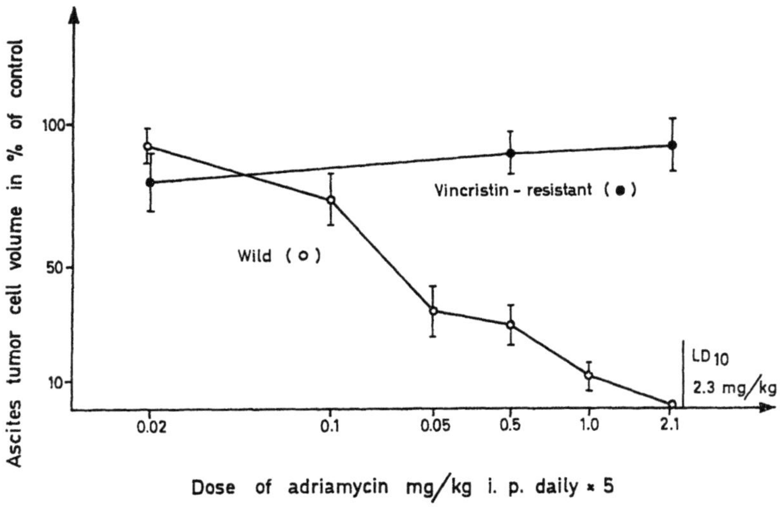

Fig. 5. Response to treatment with adriamycin of a wild Ehrlich ascites tumor and of a subline selected for resistance to vincristine. For experimental conditions and explanation of signs see text to Fig. 2. Response of the wild tumor to vincristine at 0.25 mg/kg ip daily × 5 was inhibition to 67% ± 7% (mean value ± SE) of control while the vincristine-resistant tumor with the same dose grew to 111% ± 10% of control

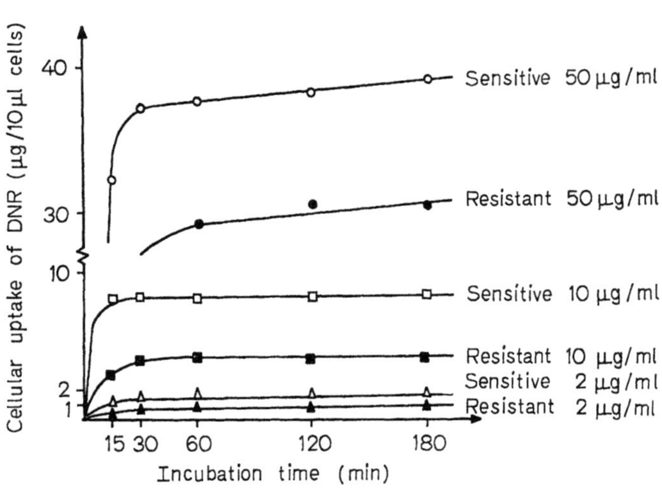

Fig. 6. Cellular uptake of daunomycin in sensitive and resistant Ehrlich ascites tumor cells in vitro. 1% cell suspensions were incubated at 37% and daunomycin added at time zero to give the concentrations indicated. Daunomycin remaining in the medium and in 2 batches of washfluid and daunomycin extracted from the cells with phenol/chloroform/SDS was determined spectrophotometrically, and the amount of drug taken up was calculated

found *in vitro*, as seen in Fig. 6, showing the uptake in sensitive and resistant cells with different initial concentrations of daunomycin. Investigations on inhibition of nucleic acid synthesis with daunomycin in sensitive and resistant cells in vitro [Danø,

FREDERIKSEN and HELLUNG-LARSEN: Cancer Res. **32**. 1307-1314 (1972)] showed a difference in degree of inhibition, which was too great to be explained by the difference in uptake, indicating that other factors also play an important role in the development of resistance. Possible factors are a directly altered sensitivity of nucleic acid synthesizing systems, or alterations in intracellular drug distribution or drug metabolism. Preliminary experiments show that decreased cellular drug uptake is also a factor in the mechanism of adriamycin resistance.

The reason for the cross-resistance between daunomycin and adriamycin, and between these drugs and vincristine, remains to be investigated. It may be due to similarities in the mechanisms of cellular uptake of the drugs, similarities in drug binding or drug metabolism, or similarities in their mechanism of action.

Part IV:

Pharmacology and Pharmacokinetics

Influence of Pharmacokinetic Variations on the Pharmacological Properties of Adriamycin[1]

E. ARENA, N. D'ALESSANDRO, L. DUSONCHET, N. GEBBIA, F. GERBASI, R. SANGUEDOLCE, and L. RAUSA[2]

Istituto di Farmacologia dell'Universita' di Palermo (Direttore: Prof. P. BENIGNO)

With 22 Figures

Introduction

Whenever it appears impossible to modify the chemical structure of drugs with a high and established therapeutic activity but a low chemotherapeutic index, pharmacological research has to find other ways of improving the chemotherapeutic index. This problem is particularly important in the case of antitumor drugs, thus justifying research into the most suitable choice of dosage and routes of administration, as well as into the pharmacological associations which enable tumor cells to be hit at various stages of the reproductive cycle. Alternatively, the therapeutic index could be improved by the use of antagonistic compounds (like, for example, methotrexate and folinic acid) which act upon the same organic functions.

The latter approach, which modifies the action of drugs on action receptors in a relatively controllable way, can also be exploited through the simultaneous administration of drugs which, though having no obvious pharmacodynamic affinity, can modify the pharmacokinetics of a chemotherapeutic drug through their combined action on transport or storage receptors.

The variations in drug distribution caused by this process will be regarded as favorable or unfavorable, according to their effect on the chemotherapeutic index. We have observed such interactions in the course of studies on modifications of the acute toxicity of daunomycin [1]. For instance, pretreatment of animals with tetracycline or acetylsalicylic acid „*per os*“ significantly reduces or increases the mortality caused by daunomycin administered intravenous; on the other hand, the modifications caused by pretreatment with sulfamethoxypyridazine seem relatively unimportant.

---

[1] Research supported by C.N.R. (National Research Council), Committee for Antibiotic Toxicity, Contract no. 70. 00714/31/19.9.3.

[2] Associate Professor of Pharmacology and Medical Therapy, University of Palermo.

Though it is not easy to show the existence of a clear relationship between pharmacokinetic and activity variations through the use of such a rough index as LD_{50}, we have explained this phenomenon on the basis of a modification in the distribution of the drug which, by increasing or decreasing the concentration of the chemotherapeutic agent in some organs, strongly influences its interaction with the action receptors. After pretreatment with each of the three above-mentioned compounds, the antitumor-drug concentration increases in the blood and falls in all other parenchymas except the spleen (in the case of pretreatment with sulfonamide) and the kidney (in the case of pretraetment with tetracycline). The problem with adriamycin was how to improve its therapeutic index. Since it can be characterized and evaluated spectrofluorimetrically by a technique which identifies the anthracyclinic group, we set out to study the pharmacokinetic characteristics of adriamycin in an attempt to modify its distribution in the heart, first reports on the clinical use of adriamycin having indicated a certain cardiotoxic activity which, although not very high, could limit the use of the drug in human therapy.

In addition, our *"in vivo"* and *"in vitro"* studies were aimed at investigating, in particular, the potential cardiotoxic activity of adriamycin, the most likely mechanism of action of this cardiotoxic activity and the possible ways of modifying it through the administration of various drugs.

Materials and Methods

A. *"In vivo" Experiments*

1. Pharmacokinetic Experiments

Adriamycin distribution in the body was studied on groups of 10 COBS mice into which Sarcoma 180 had been transplanted 10 days before. Adriamycin was dissolved in saline solution and injected i. v. at the dose of 8 mg/kg. The evaluation of the antitumor drug in tissues was carried out 10 min, and 4, 8, 12, 24, 48 and 72 h after its administration.

2. Autohistographic Study of Distribution

The experiments were carried out on COBS mice bearing Sarcoma 180 and treated with a dose of 10 mg/kg i. v., using the technique described by TUBARO and BULGINI [3].

3. Evaluation of Biliary and Renal Excretion.

Adriamycin and daunomycin were administered i. v. at 2 mg/kg through the marginal vein of the ear to rabbits after 24 h of fasting. The concentration of the two drugs at different times was measured in the bile, obtained by cannulation of the common bile duct, and in the urine, obtained by catheterization.

4. Study on the Existence of an Enterohepatic Circulation

Doses of 20 mg/kg of adriamycin were administered to rabbits *"per os"*, following neutralization of the gastric juice, and the presence of the drug in blood and bile was investigated. In addition, two groups of 5 rabbits each received 4 mg/kg of adriamycin for 6 successive days; the drugs was administered *"per os"* to one group and intravenously to the other group. The number of dead animals and their gastrointestinal symptoms were recorded.

5. Influence of the Treatment on Spleen and Heart Concentrations of the Antiblastic Drug

a) Repeated Doses: 2.7 mg/kg of adriamycin were administered intravenously for 3 days to a group of COBS mice carrying Sarcoma 180. The amount of the drug was measured in the heart and spleen of the animals, sacrificed 4 and 24 h after the last administration of the drug.

b) Subdivided Doses: One group of animals received 2.7 mg/kg of the drug i. p. daily for 3 days. A second group received 1.35 mg/kg of the drug every 12 h for 3 days i. p., so that all received the same final dose. Four and 24 h after the last administration the concentration of the drug was measured in the heart and spleen of the two groups of mice.

6. Influence of Strophanthin on Adriamycin Storage in Isolated Hearts of Rabbits and in Heart and Spleen of Rats

a) Isolated Rabbit's Heart: Strophanthin, at a concentration of 12.5 mcg/1 was added to Locke's fluid for the perfusion of isolated hearts of rabbits and perfusion was carried out so that 0.125 mcg/g were given to the isolated heart. Adriamycin was then added to the perfusion fluid up to a concentration of 1 mcg/ml. The experiment was interrupted when the myocardium had been perfused by a total amount of 10 mcg/g of the antiblastic drug.

b) Experiments on Rats: 10 mcg/kg of strophanthin were administered s. c. to groups of 10 rats daily for 3 days, 1 h before the administration of 2.5 mg/kg of adriamycin i. v. The animals were sacrificed 8 h after the last administration.

7. Identification and Estimation of Adriamycin and Daunomycin

Adriamycin or daunomycin were extracted from tissues or organic fluids by a procedure previously described [4, 5] which allows the spectrophotofluorimetric identification and evaluation of the material extracted. A Hitachi Perkin Elmer MPF 2A spectrophotofluorimeter was used for the evaluation of the drugs in the tissues.

8. Influence of Dosage-Schedule and of Strophanthin on the Chemotherapeutic Activity of Adriamycin and Daunomycin

a) Treatment by a Single Daily Dose: One day after tumor implantation on the subaxillary cavum, 2 groups of 10 COBS mice began to receive 5 mg/kg of daunomycin or 2.5 mg/kg of adriamycin daily for 8 days by subcutaneous injection. Anoother group was treated as follows: 10 mcg/kg of strophanthin by intramuscular injection and, 1 h later, 2.5 mg/kg of adriamycin s. c. The treatment was carried out daily, for 8 days. On the 10th day after implantation the spleen and tumor weights were assessed for the 3 groups of mice.

b) Subdivided Treatment: Following the same technique as in the case of the single daily treatment schedule, daunomycin and adriamycin were administered s. c., at doses of 2.5 mg/kg and 1.25 mg/kg, respectively, every 12 h. Spleen and tumor weights were recorded on the 10th day.

9. Influence of Strophanthin on the Mortality Caused by Adriamycin

A group of 20 COBS mice was treated with 12 mg/kg of adriamycin i. p. A second group received 16 mcg/kg of strophanthin, i. m., 1 h before the administration of adriamycin.

10. Effects of Adriamycin on the Cardiovascular Apparatus.

a) Experiments on Rabbits: The action of adriamycin was evaluated on a „pressure and breathing" rabbit preparation by rapid intravenous injection of 20 mg/kg. Electrocardiographic tracings were obtained from animals to which 20 mg/kg of adriamycin or daunomycin had been administered intravenously. The ECGs reproduced in the figures were obtained by thoracic derivation, according to Block.

b) Experiments on dogs: Groups of 5 dogs, were intubated and ventilated in a closed circuit through a Palmer pump. After thoracotomy, transducers were applied for flow measurements, one on the aorta and another on the circumflex branch of the left coronary artery after pericardiotomy. Both transducers were then connected to an electromagnetic Statham flowmeter. The arterial blood pressure was measured with an electromagnetic apparatus after the introduction of a catheter into the femoral artery. The force of contraction of the myorcadium, expressed in grams, was determined by means of a strain-gauge sutured to the free wall of the left ventricle. Doses of up to 4 mg/kg adriamycin were administered to these animals. β-carbobenzyloxy-aminomethyl-1,6-dimethyl-10-ergoline maleate (MCE) (a substance exerting an anti-serotoninic action) was injected i. v. at the dose of 1.5 mg/kg into 4 other animals, a few minutes before the administration of 1.5 mg/kg of the antiblastic drugs under test.

11. Test of the Catecholamine Content in Rat's Heart and Suprarenal Capsules

The animals received 2.5 mg/kg of the drug i. p. every 12 h for 4 days and were sacrificed 1 h after the last administration. The catecholamine content was measured according to Giotti and Jngianna [8].

B. "*In vitro*" *experiments*

1. Experiments on Isolated Rabbit's Heart

Hearts of rabbits, killed by fracture of the cervical vertebrae, were removed and their aorta rapidly connected to the polyethylene cannula of a Spadolini apparatus for the perfusion of the isolated heart. Locke's fluid was used for perfusion and the drugs under test were added at the concentrations shown in the corresponding figures.

2. Study of the Influence Exerted by Adriamycin and Daunomycin on the Incorporation of Tritium-Labelled Thymidine and Leucine by Isolated Cells of sarcoma 180

Cells of sarcoma 180 were isolated by trypsinization of the solid tumor and suspended in order to obtain 25 mg (f. w.) in a final volume of 1.5 ml of Hank's medium containing: lactalbumin 5%, calf extract 5%, calf serum, $NaHCO_3$ 4.4%, penicillin 200 I.U./ml, streptomycin 200 mcg/ml, 10 μCi of the tracer employed, i. e. ^{3}H-6-thymidine (2 Ci/mM) or L-^{3}H-leucine (250 mCi/mM) provided by Amersham Radiochemical Centre, and daunomycin (10 mcg/ml) or adriamycin (up to 60 mcg/ml). Strophanthin up to concentrations of 60 mcg/ml was added when required by the experimental plan. The incorporation of ^{3}H-thymidine into DNA was measured according to Bollum [6], that of ^{3}H-leucine into proteins according to Mans and Novelli [7]. Radioactivity was evaluated by means of a scintillator in a liquid phase.

3. Oxygen consumption and oxidative phosphorylation

Oxygen consumption by liver and heart homogenates was evaluated according to the method described by Umbreit et al. [9]. Adriamycin and Daunomycin were added to Warburg flasks up to a concentration of 0.1 mM/ml of homogenate. Oxidative phosphorylation of mitochondria of rat liver and calf heart was studied according to Hunter [10]. Adriamycin and daunomycin were added in the Warburg flasks at a concentration of 0.1 mM/3 ml of reaction mixture.

4. Test of Na^+K^+-Stimulated, Mg^{++}-Depent ATPase Activity

The action of adriamycin or daunomycin up to a concentration of 10^{-4} M was essayed on Na^+K^+-stimulated ATPase, present in the particulate fraction (membranes) sedimented at 10000 x g according to the method of MATSUI and SCHWARTZ [11]. The amount of proteins in the reaction mixture was kept at 250 mcg. The inorganic phosphate was measured according to FISKE and SUBBAROW [12].

5. CPK Activity

The action of adriamycin and daunomycin on CPK was tested on a pure preparation of this enzyme supplied by Boehringer. The reaction mixture contained 10 mU of enzyme and 5 mcg/ml of the antiblastic drug.

6. Superprecipitation of Actomyosin

Actomyosin was extracted from the skeletal muscle of rabbits, according to EBASHI et al. [13]. Adriamycin and daunomycin were added to the sample at a concentration of 10^{-4}M before the addition of Ca^{++} and ATP. The ATPase activity of actomyosin was determined at the same time.

7. ^{45}Ca Uptake by Calf-Heart Sarcotubules and Mitochondria.

Sarcotubules and mitochondria were prepared according to the technique of FANBURG [14]. The rate of $^{45}CaCl_2$ uptake by sarcotubules was followed for 40 min at 22°C by incubating this fraction in a medium containing: histidine 20 mM (pH 7.4), $MgCl_2$ 5mM, ATP 5mM, $^{45}CaCl_2$ up to 0.6 mM. The protein concentration was 2.8 mg/ml. Some trials were carried out in the presence of potassium oxalate 5 mM. Adriamycin and daunomycin were added at a concentration of 10^{-4}M. In the trials performed with mitochondria incubated in the same buffer, $^{45}CaCl_2$ was kept at a concentration of 0.4 mM. The amount of mitochondrial proteins was 4 mg/ml. The antiblastic drugs were added up to a concentration of 10^{-2}M.

Experimental Results

Fig. 1 shows the pharmacokinetic characteristics of adriamycin in COBS mice bearing sarcoma 180, after treatment with 8 mg/kg of this substance i. v. As we also

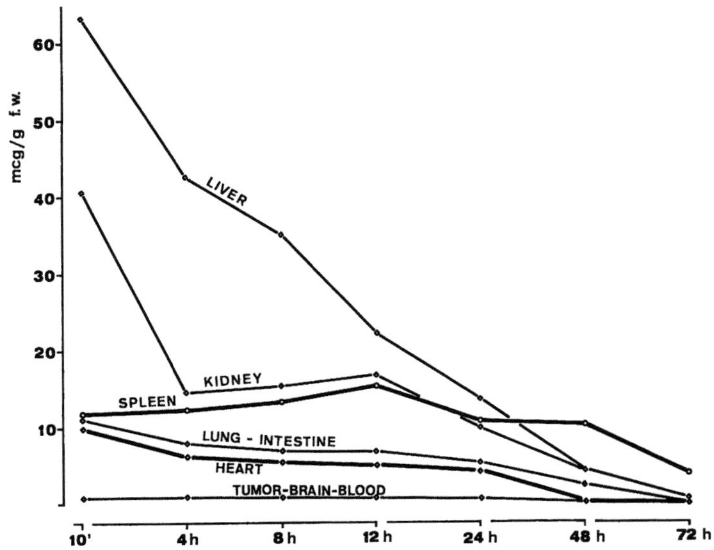

Fig. 1. Concentration of adriamycin in tissues of COBS mice carrying sarcoma 180 and treated with a dose of 8 mg/kg i.v.

observed with daunomycin [1], adriamycin is readily transferred from the blood to the tissues where it reaches remarkable concentrations in all the organs except the tumor and the brain. The drug is still present in the liver and kidneys 72 h after the treatment; it disappears from the lungs and the intestine between 48 h and 72 h.

The concentration increase in the spleen between 4 and 12 h is worthy of note. It should also be pointed out that the highest concentration of the antitumor drug, at 72 h, is found in the spleen; this pharmacokinetic property has also been noted by other authors, employing the labelled molecule (15). In the heart, the drug quickly reaches high levels; 50% of the concentration found 10 min after the treatment is still present at 24 h.

An examination of the distribution map in COBS mice carried out with an auto-histographic technique [3] enabled us to confirm the results we had obtained with the spectrofluorimetric method, to find high concentrations of the drug in the sub-maxillary gland (Fig. 2).

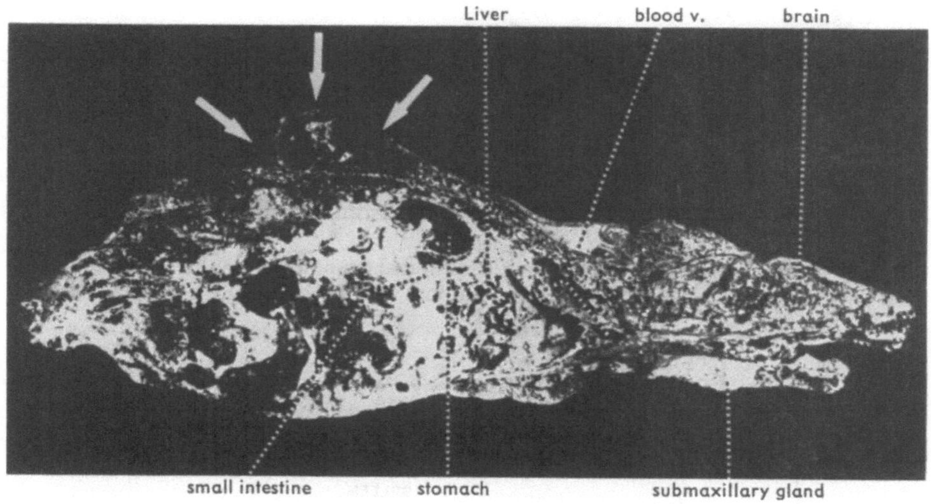

Fig. 2. Sagittal section (200 microns) of a COBS mouse carrying sarcoma 180 treated with a dose of 10 mg/kg of adriamycin by intravenous route and sacrificed 10 min after the administration of the drug. The white arrows indicate the presence of adriamycin

Tests to detect biliary and urinary excretion were carried out on rabbits by comparing the excretion rate of adriamycin with that of daunomycin (Fig. 3). The results show that at 5 h about 12% of administered adriamycin is eliminated from the body of the experimental animals; the excretion rate of daunomycin is twice that of adriamycin. There is practically no urinary excretion of either drug.

The experiments carried out on rabbits to investigate the possible existence of an enterohepatic circulation showed that adriamycin, like daunomycin, is not absorbed by the gastroenteric tract; when 100% lethal (i. v.) doses were administered "*per os*" the rabbits showed no signs of toxicity.

The histophilic properties of adriamycin, besides being responsible for the long half-life of the drug in the tissues, also cause the storage phenomena we were able

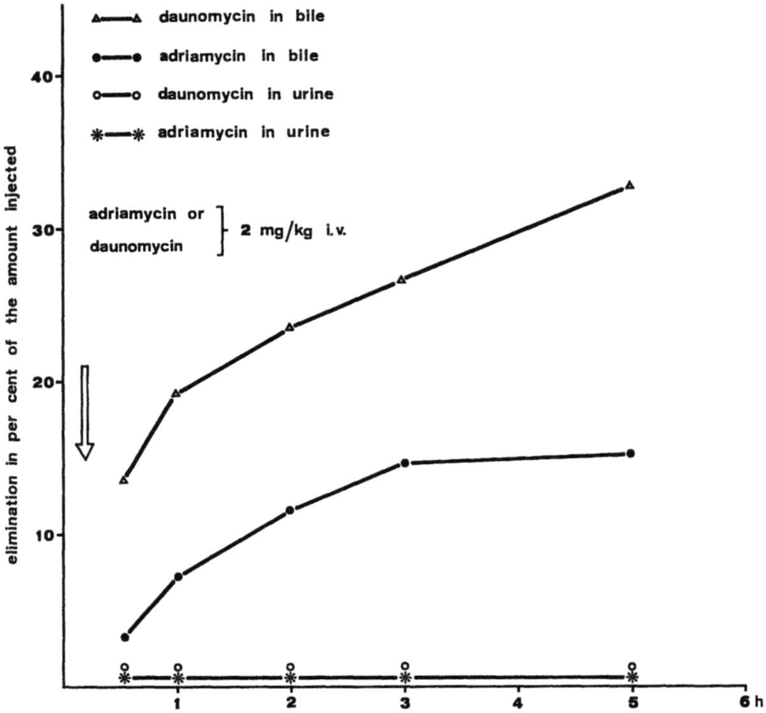

Fig. 3. Biliary and urinary excretion of adriamycin and daunomycin in the rabbit, following the administration of a dose of 2 mg/kg i.v.

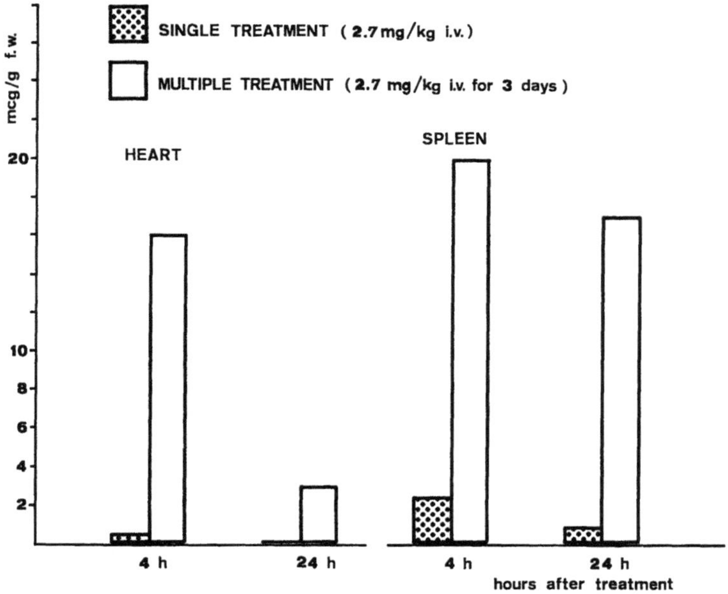

Fig. 4. Levels of adriamycin in the heart and spleen of COBS mice treated i.v. with adriamycin (single or treatment repeated for 3 days)

to demonstrate by treating COBS mice with 2.7 mg/kg of the drug i. v. for 3 days and measuring the concentration of the drug in the spleen and in the heart 4 and 24 h after the last dose (Fig. 4).

The same treatment plan as described for Fig. 4 was followed, but the 2.7 mg/kg daily amount was divided into two doses of 1.35 mg/kg administered every 12 h i. p. (Fig. 5). On day 3 of treatment, 4 and 24 h after the last administration, the concentration of the drug in the heart of animals treated with subdivided doses was lower than that in animals which had received the single daily dose.

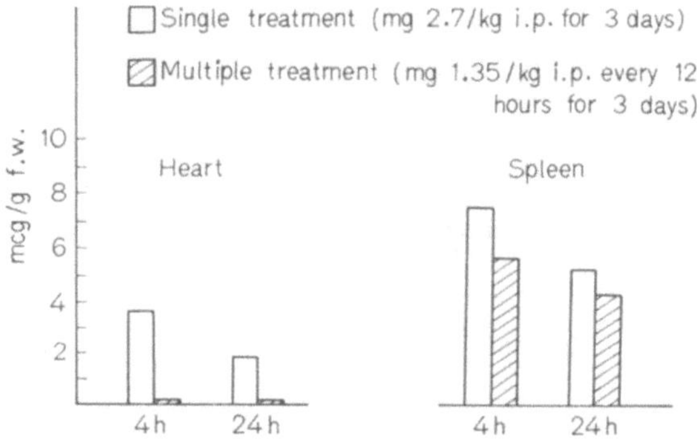

Fig. 5. Levels of adriamycin in the heart and spleen of COBS mice treated i.p. with adriamycin (single of subdivided doses for 3 days)

The spleen did not show the same pharmacokinetic behaviour, the differences in the results obtained with the two treatments being negligible.

Treatment of COBS mice with equiactive doses of the two drugs was started the day after inoculation of the tumor and carried out for 8 days, according to the dosage schedule described by Di Marco et al. [16]. Adriamycin exhibited a higher activity than daunomycin (Fig. 6). When half the equiactive dose was administered every

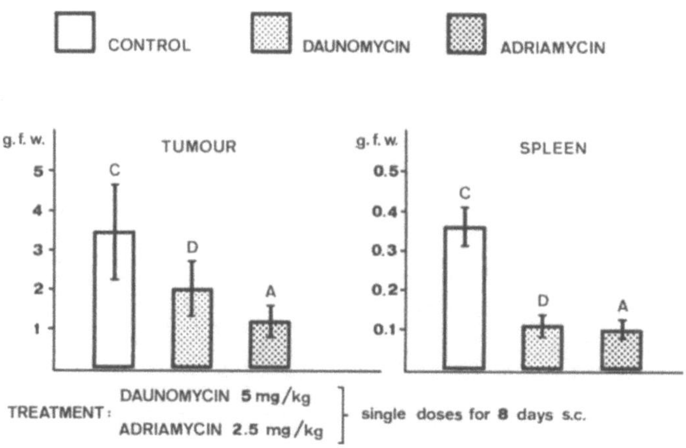

Fig. 6. Antineoplastic activity of adriamycin and daunomycin in COBS mice bearing sarcoma 180 (single doses)

12 h, no appreciable changes in the antineoplastic activity of adriamycin were observed (the weight of tumors being used to measure antineoplastic activity) while the activity of daunomycin practically doubled (Fig. 7).

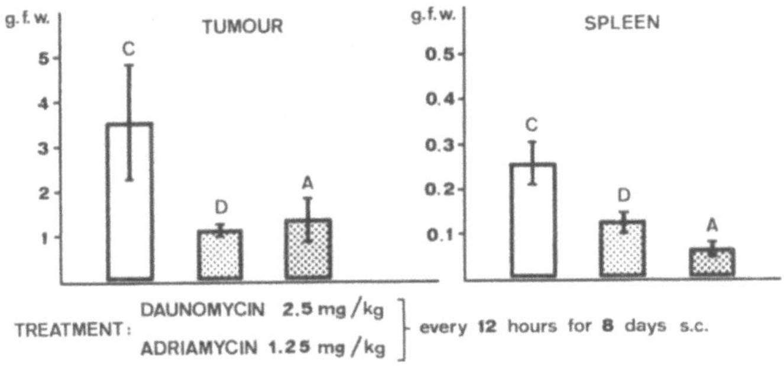

Fig. 7. Antineoplastic activity of adriamycin and daunomycin in COBS mice bearing sarcoma 180 (divided doses)

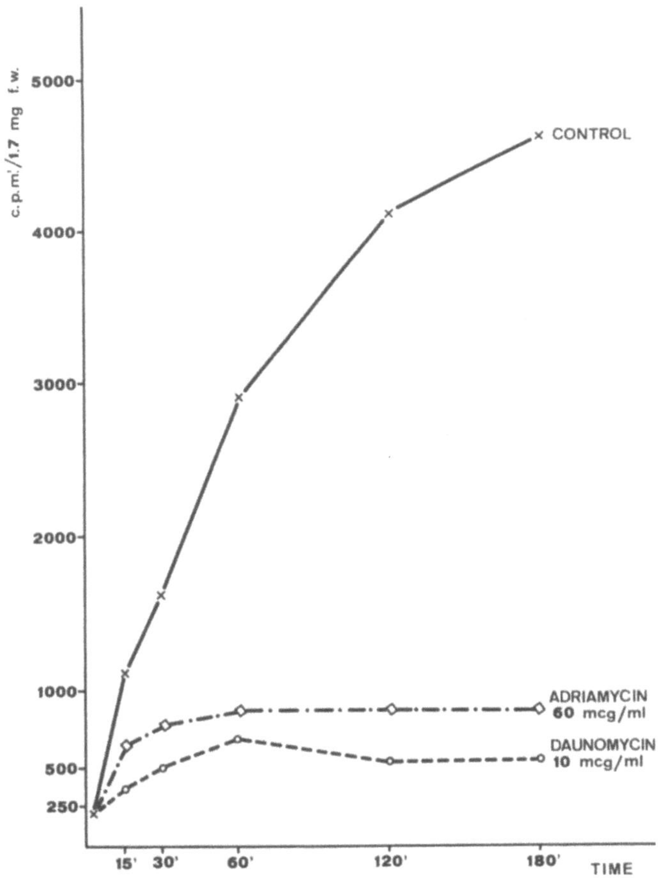

Fig. 8. Incorporation of tritium-labelled thymidine into isolated cells of sarcoma 180, in the presence of adriamycin and daunomycin

In order to explain the higher antineoplastic activity of adriamycin, two series of experiments were carried out. In the first series of trials, COBS mice carrying sarcoma 180 were treated for 3 days with equal doses of both drugs (2.7 mg/kg i. v.), in order to study the relationship between drug concentration in the tumor mass and antineoplastic effect. The animals were sacrificed 8 h after the last administration. Titration of the two drugs in the tumor showed that the concentration of the two antiblastic drugs was the same (1.2 mcg/g f. w.).

In the second series of trials, the two substances were added to isolated cells of sarcoma 180, in the presence of labelled thymidine and leucine (Fig. 8 and 9). The graph shows the lower "*in vitro*" activity of adriamycin on thymidine incorporation, as compared to daunomycin. In fact, 60 mcg/ml of adriamycin were added to the substrate before that the same inhibition activity was observed with 10 mcg/ml of daunomycin (Fig. 8 and 9). This phenomenon has been also reported by other authors (17-18). The greater effectiveness of the drug "*in vivo*" could be explained by an activation process of adriamycin.

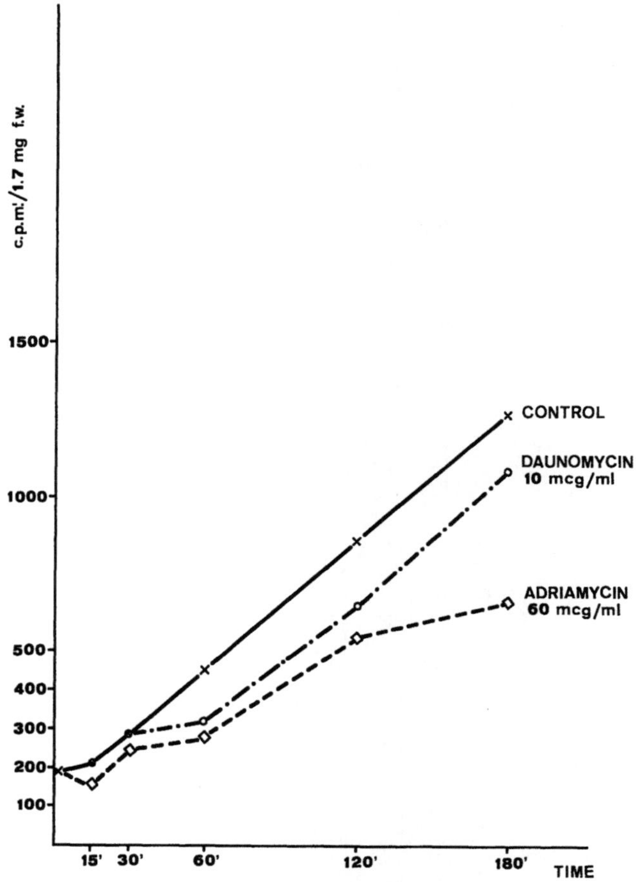

Fig. 9. Incorporation of tritium-labelled leucine into isolated cells of sarcoma 180, in the presence of adriamycin and daunomycin

This hypothesis is supported by the results of an experiment consisting in the administration of equal doses of the two antimitotic drugs to mice carrying sarcoma, for a period of 3 days.

In fact, since equal concentrations of both drugs were found in the tumoral mass, the different antitumor activity of the two drugs "*in vivo*" cannot be attributed to a higher level adriamycin in the tumor mass.

In a further analysis of other pharmacodynamic actions of adriamycin performed with a "pressure and breathing" rabbit preparation (Fig. 10), no modifications of

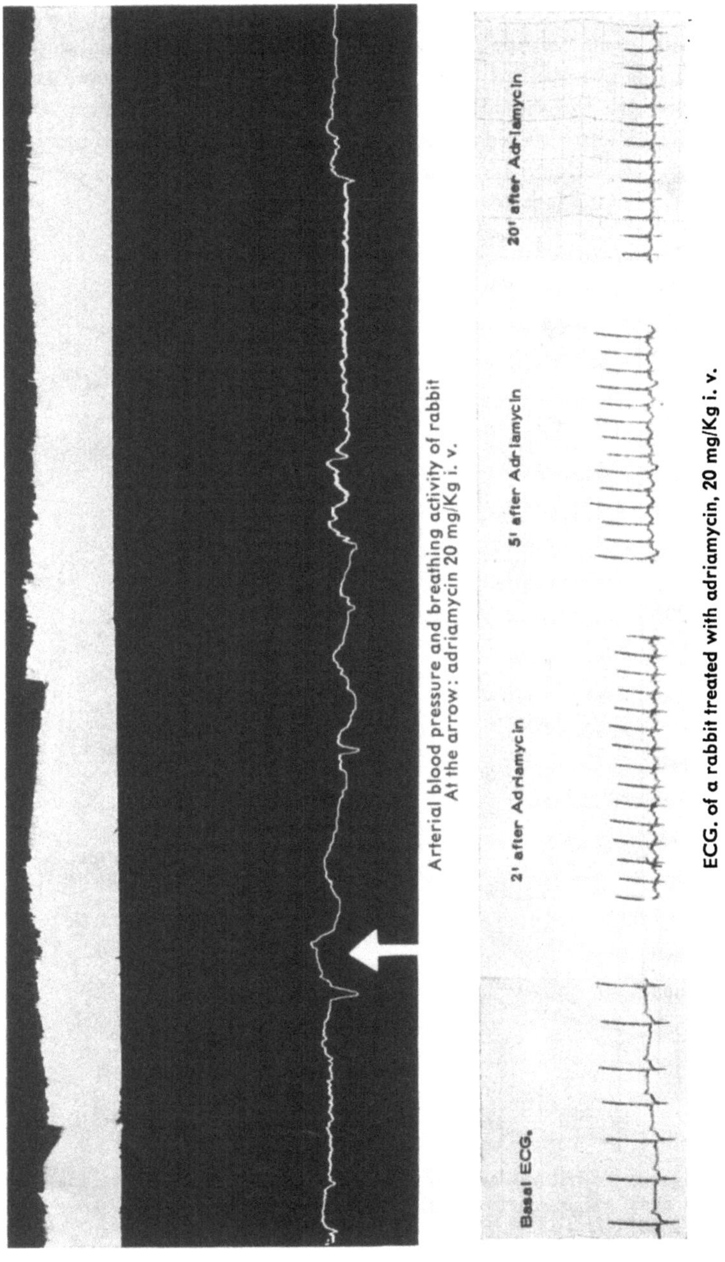

Fig. 10. Rabbit. Effect of adriamycin on systemic arterial pressure, breathing and ECG

the pressure and breathing values were observed following rapid injection of the drug at the dose of 20 mg/kg i. v., but the ECGs of 50% of the animals showed rapid onset of sinusal tachycardia associated with a T-wave inversion.

This ECG does not show the abrupt rhythm variations which are observed in a high percentage of rabbits treated with equal doses of daunomycin (Fig. 11).

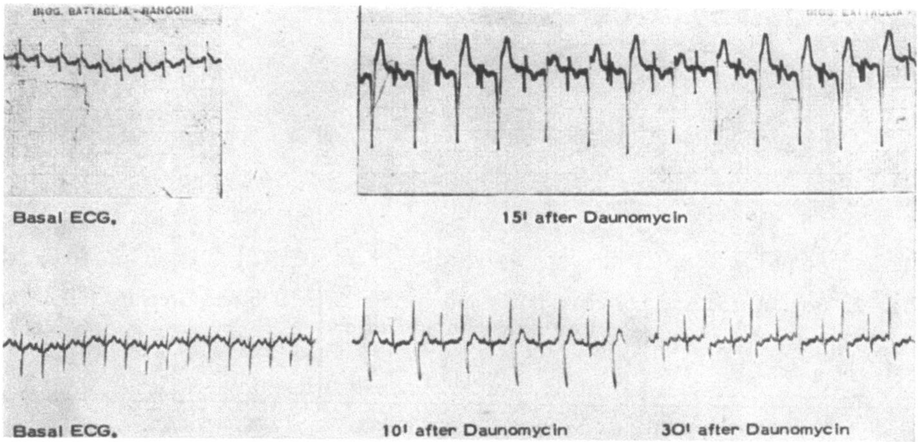

ECGs. of rabbits treated with adriamycin, 20 mg/Kg i. v.

Fig. 11. Rabbit. ECG following treatment with daunomycin

To acquire further knowledge about the effects of the drug on the heart, the above research was extended to dogs and the following parameters were analyzed: inotropism by suture of a straingauge in the myocardium; coronary and aortic flows ba electromagnetic flowmeters; arterial blood pressure; ECG. It was thus possible to observe that 500 mcg/kg of adriamycin, injected intravenously, give a moderate and temporary positive inotropic effect.

With a drug dose of 1.5 mg/kg, inotropism, coronary flow, aortic flow and arterial blood pressure show a temporary increase (Fig. 12 and Table 1). Six minutes after treatment the coronary flow returns to its initial values and then gradually decreases until it reaches a minimum at 15 min (−30%).

The same phenomenon is observed with regard to inotropism (Fig. 13).

At the same time in the ECG tracings (not reported here) a moderate sinusal tachycardia was noted immediately after the administration of the drug. The effects of adriamycin on the force of contraction of the myocardium resemble those effects which, at least theoretically, could be antagonized by the digitalic glycosides. Fig. 13 shows the action of strophanthin which, at the dose of 16 mcg/kg, quickly normalizes inotropism and coronary flow; at the same time, the aortic flow tends to stabilize at a level higher than the initial one. Fig. 14 shows the tracings of the aortic and coronary flows in a dog, 40 min after administration of 1.5 mg/kg of adriamycin. The typical depression caused by the antiblastic can be observed. As can be seen, the "restitutio ad integrum" induced by the cardiac glycoside is extremely rapid.

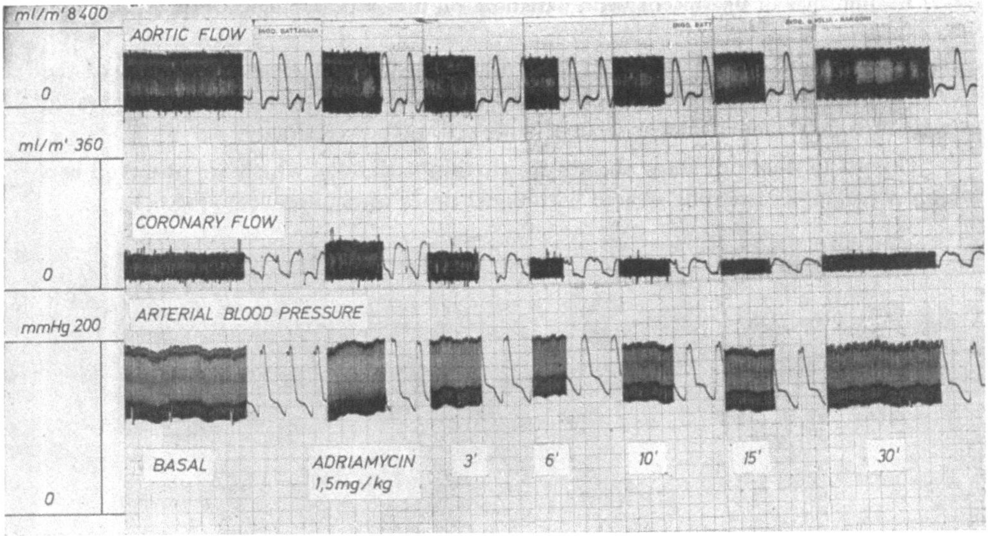

Fig. 12. Dog. Aortic and coronary flows and femoral arterial blood pressure following treatment with adriamycin

Table 1

Drugs	Mean blood pressure in mm Hg registered from femoral artery		Aortic flow Maximum ml/min	Aortic flow Mean ml/min	Coronary flow Maximum ml/min	Coronary flow Mean and ml/min	%variation
Adriamycin 1,5 mg/kg	Basal	130	5480	1400	93	51	
	After infusion	132	5480	1680	123	57,6	+13%
	After 3 min	142	5200	2200	93	62	+21%
	After 6 min	136	4680	1940	72	52	+12%
	After 10 min	141	4640	980	62	43	−16%
	After 15 min	138	4780	990	52	36	−30%
	After 30 min	130	4616	980	58	35	−31%
Administration of strophanthin 16 μg/kg	Basal	130	4480	980	57	36	
	After infusion	150	4760	2520	93	43	+19%
	After 5 min	158	3320	2580	94	57	+58%
	After 10 min	157	5400	2390	100	58	+59%
	After 15 min	157	5480	2680	100	58	+59%

	Force of systolic contraction	g	%variation
Adriamycin 1,5 mg/kg	Basal	27,5	
	After infusion	30	+ 9.1%
	After 3 min	30	+ 9.1%
	After 6 min	23	−16.4%
	After 10 min	22	−20%
	After 15 min	19	−31%
	After 30 min	18	−34%
Administration of strophanthin 16 μg/kg	Basal	18	
	After infusion	25	+33%
	After 5 min	25	+33%
	After 10 min	27	+35%
	After 15 min	27	+35%

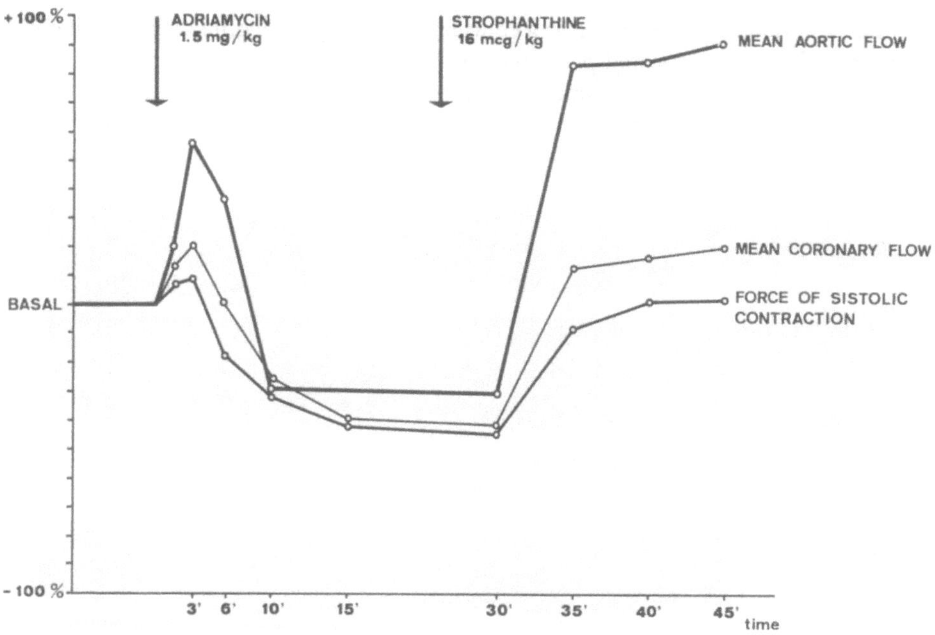

Fig. 13. Variations of the coronary and aortic flows and of inotropism in a dog treated with 1.5 mg/kg of adriamycin and the with 16 mcg/kg of strophanthin

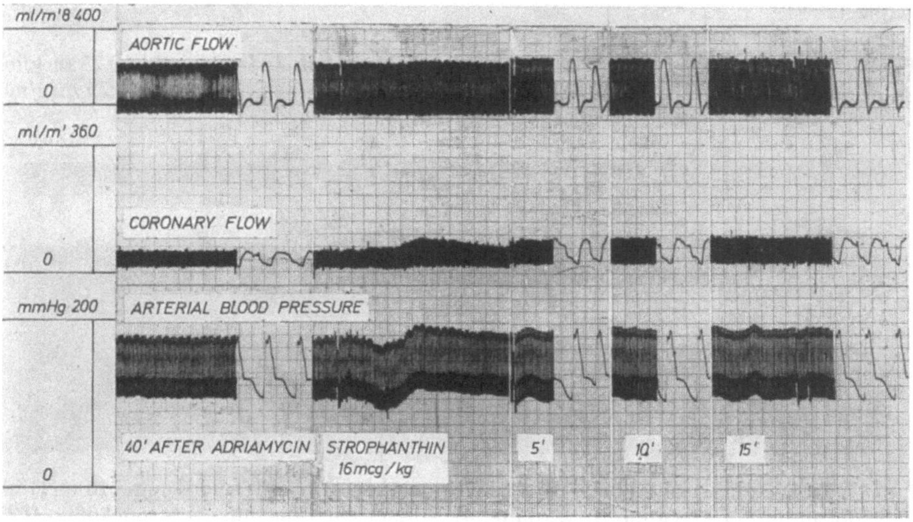

Fig. 14. Effect of strophanthin on the aortic and coronary flows and on the femoral arterial blood pressure of a dog treated with 1.5 mg/kg of adriamycin

In order to analyze the factors which could be responsible for the toxic effect of adriamycin an the heart, the action of the antitumor drug was tested on isolated hearts of rabbits. Fig. 15 shows a quick lowering of inotropism, which takes place when 1 mcg/ml of drug is added to the isolated organ.

The addition of strophanthin at a concentration of 0.25 mcg/ml after washing did not affect the above-reported phenomenon.

Fig. 16 shows how the same negative inotropic action is exhibited by adriamycinone. Strophanthin added to the perfusion fluid up to a concentration of 0.25 mcg/ml seems to antagonize the action of the aglycone, but only to a slight extent.

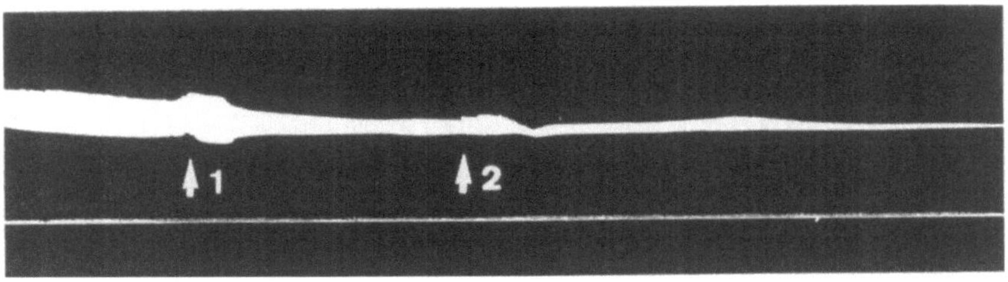

Fig. 15. Isolated heart of rabbit after addition of adriamycin and then of strophanthin to the perfusion fluid

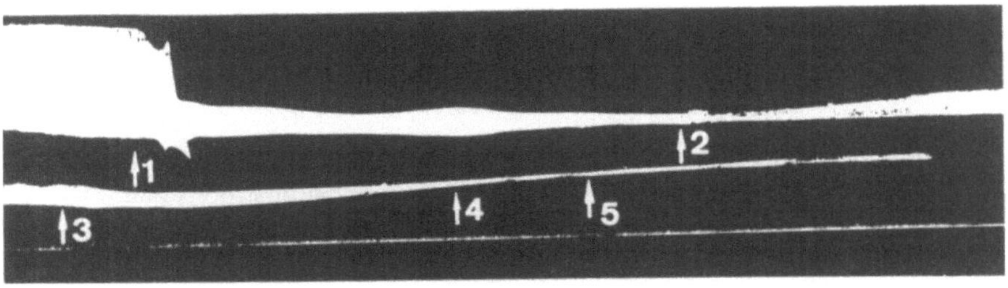

Fig. 16. Isolated heart of rabbit. 1) Adriamycinone 1 mcg/ml, 2) Strophanthin 0.25 mcg/ml + Adriamycinone 1 mcg/ml, 3) Adriamycinone 1 mcg/ml, 4) Strophanthin 0.25 mcg/ml, 5) Washing

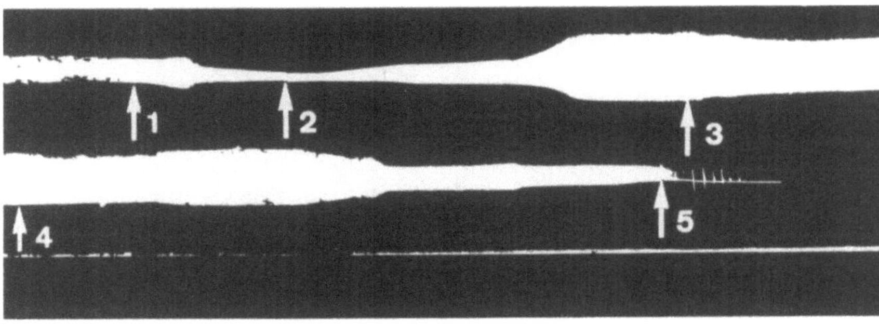

Fig. 17. Isolated heart of rabbit. 1) Adriamycinone 1 mcg/ml + Daunosamine 10 mcg/ml, 2) Strophanthin 0.25 mcg/ml, 3) Washing, 4) Adriamycin 1 mcg/ml, 5) Strophanthin 0.25 mcg/ml

However, when the heart is perfused again, after washing, with 1 mcg/ml oft he drug, the effect becomes irreversible and is no longer antagonized by strophanthin or by washing.

Carrying out perfusion of the heart with a mixture of the two components of adriamycin, i. e. adriamycinone and daunosamine (aminoglycide which we have found to have no activity on the heart) mixed at a non-equimolecular ratio, it was possible to observe the classical negative inotropic effect of the aglycone. Under these experimental conditions, strophanthine can normalize the force of contraction of the myocardium, and adriamycin added after the washing process does not cause negative inotropic actions, probably because some strophanthin remains bound to the heart tissues. In any case, at a certain time after perfusion with adriamycin, inotropism starts to decrease and the addition of relatively high concentrations of strophanthin only seems to make the situation worse.

Pretreatment of the heart with strophanthin (12.5 mcg/l) does not affect inotropism, not even in the case of a concentration of adriamycin 20 times higher than that capable of causing cardiac arrest (Fig. 18). When, after washing, perfusion is carried out with 1 mcg/ml, a certain negative inotropic effect takes place and subsequently becomes stronger, probably because the residual strophanthin leaves the myocardium.

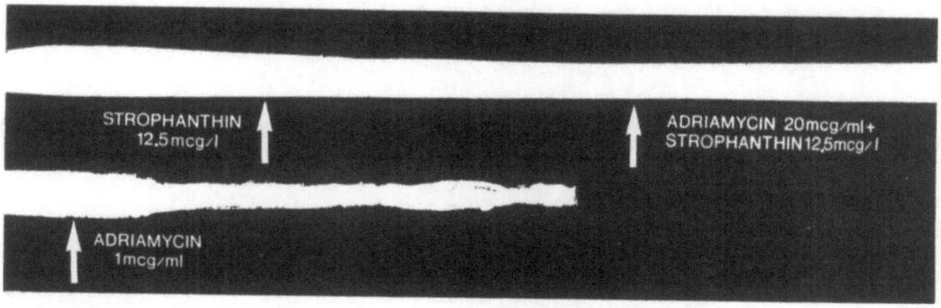

Fig. 18. Isolated heart of rabbit following the addition of strophanthin and, subsequently, adriamycin to the perfusion fluid

The experiments performed on the isolated organ, pretreated with a beta-blocking agent (propranolol) (Fig. 19), with an antiserotoninic substance (MCE) and with quinidine did not evidence any antagonistic action; on the contrary, the negative inotropic effect of these three agents was found to be additive to the effect caused by the antiblastic drug.

To explain this phenomenon, two series of trials were performed. The first sought to find possible pharmacokinetic mechanisms involved in the phenomenon, the second possible biochemical or biophysical causes. In the course of the first series of experiments it was observed that, when an isolated rabbit heart is perfused with adriamycin, high concentrations of the drug are found in the organ. Simultaneous perfusion with strophanthin and adriamycin clearly reduces the ability of the myocardium to bind the antiblastic drug.

Other researches let us to rule out a chemical antagonistic process, so we focused our attention on an "*in vivo*" study of the modifications caused by strophanthin on the kinetics of adriamycin in the rat. Fig. 20 shows the level of adriamycin in the heart and spleen of animals treated for 3 consecutive days, and the concentrations of adriamycin in the heart of animals injected with strophanthin 1 h before administration of adriamycin. The same kinetic changes previously observed „*in vitro*“ were also observed "*in vivo*". In fact, the concentrations found in the heart of the animals

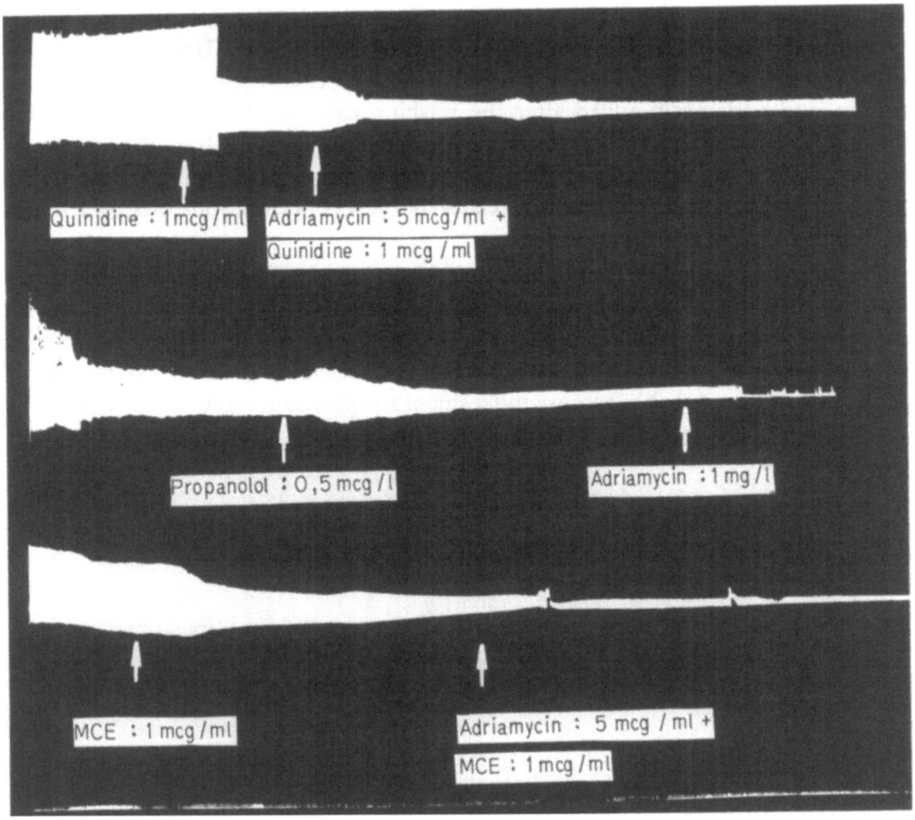

Fig. 19. Isolated heart of rabbit. Action of adriamycin after pretreatment with quinidine, propranolol and MCE (antiserotoninic substance)

pretreated with strophanthin were much lower than those of control animals, whereas the antiblastic concentrations in the spleen do not show significant modifications.

When the same trials were performed on dogs, it was observed that pretreatment with 16 mcg/kg of strophanthin inhibited the appearance of toxic effects in the heart due to adriamycin, even when the dose was 4 mg/kg, i. e. about three times more than that could cause damage to the heart. This inhibition was not observed in the case of pretreatment of the animals with a synthetic antiserotoninic substance (MCE) at the dose of 1.5 mg/kg. On the basis of the above observations and in order to evaluate the influence of strophanthin on the activity of adriamycin, the toxic and antimitotic effects of the drug in the presence of the cardiac glycoside were tested.

Fig. 21 shows that pretreatment with 16 mcg/kg of strophanthin remarkably reduces the LD_{50} in COBS mice treated with 12 mg/kg of adriamycin i. p.

The experiments to assess the influence of strophanthin on the antitumor activity of adriamycin (Fig. 22) showed that treatment with 10 mcg/kg of strophanthin and 2.5 mg/kg of adriamycin for 8 days does not affect the antineoplastic effect of the drug; at the same time, the weight of the spleen of the animals treated with strophanthin + adriamycin does not differ significantly from that of the same organ in the

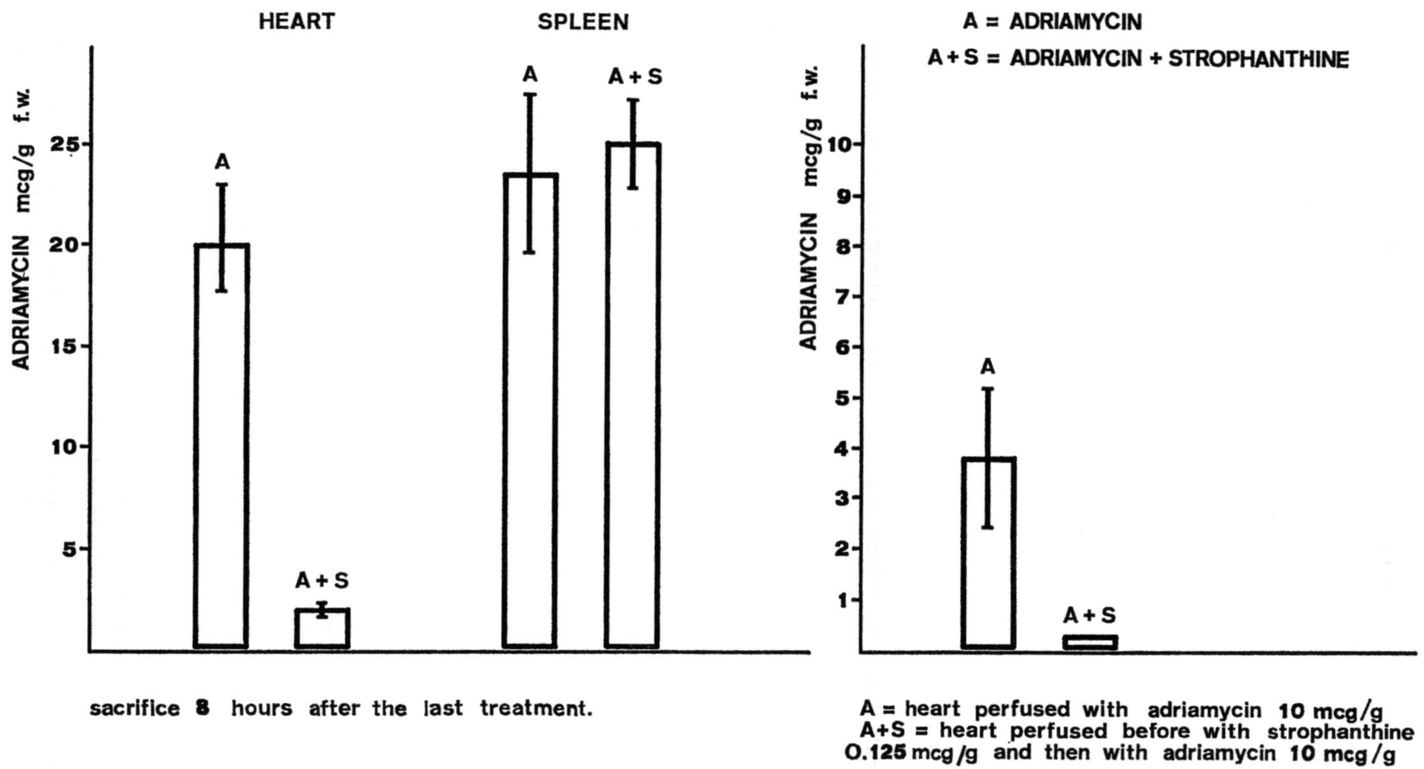

Fig. 20. Level of adriamycin in the heart and spleen of rats treated with adriamycin (2.5 mg/kg s.c.) and with adriamycin + strophanthin (10 mcg/kg i.m.) for 3 consecutive days (left) Level of adriamycin in isolated rabbit heart following treatment with adriamycin (10 mcg/g) and after treatment with adriamycin preceded by treatment with strophanthin (0.125 mcg/g) (right)

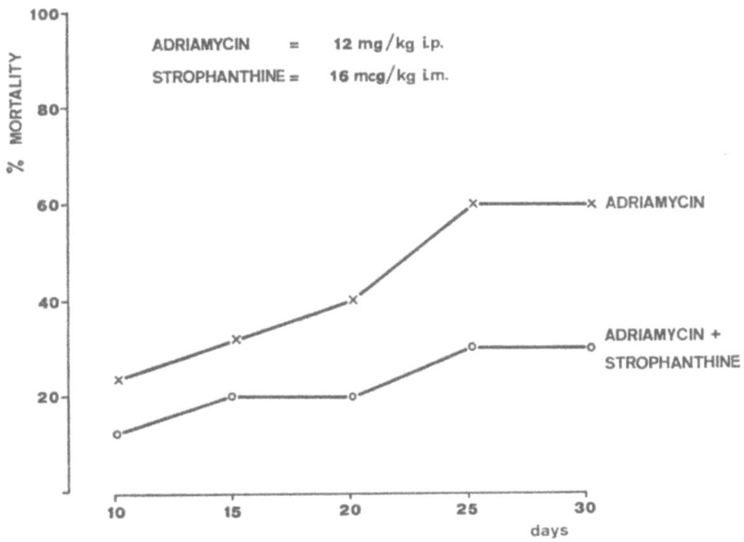

Fig. 21. Mortality caused by adriamycin and by adriamycin in combination with strophanthin

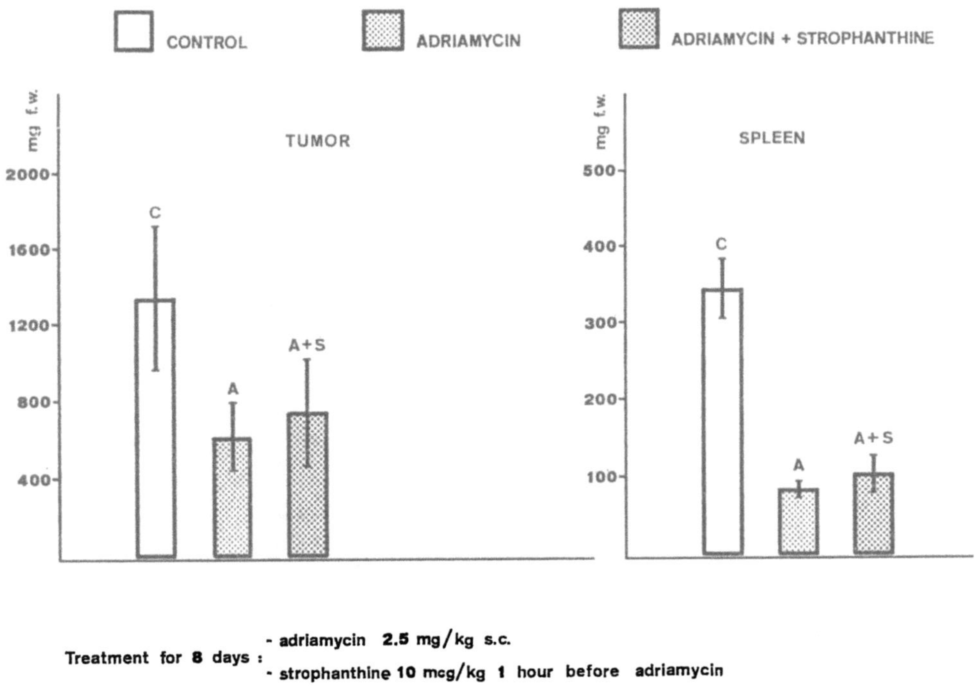

Fig. 22. Effect of treatment with adriamycin and with adriamycin + strophanthin on the weight of sarcoma 180 and of spleen in COBS mice

animals treated with adriamycin alone. These data are in agreement of the pharmacokinetic studies.

The absence of any interaction between strophanthin and adriamycin at the level of the action receptors in the tumor is further confirmed by experiments carried out on isolated sarcoma 180 cells; on this experimental model, the inhibition of ^3H-thymidine accumulation shown by adriamycin or daunomycin was not modified by the addition of up to 60 mcg/ml strophanthin.

The second series of experiments was performed in an attempt to localize the site of the process impaired by adriamycin and to investigate the antagonism to strophanthin observed on isolated fractions.

However, we were not able to detect any effect of adriamycin and daunomycin on the concentration of catecholamines in the heart and in suprarenal capsules, on O_2 consumption in liver and heart homogenates, on oxidative phosphorylation, on CPK, on the Mg^{++}-dependent, Na^+K^+-activated ATPase activity, on the supraprecipitation of actomyosin and on the uptake and release of ^{45}Ca by mitochondrial and CRS.

Conclusions

By comparing the pharmacokinetics of adriamycin with that of daunomycin, it was possible to ascertain that its greater histophilia, sustained by a high heptane-water repartition coefficient, explains the prolonged blood and tissue half-times, and the tendency to accumulation. In the combined treatment with daunomycin there was also the problem of the varying pharmacokinetics of adriamycin and consequently varying does-dependent effects in organs not involved in the therapeutic action. The results did not confirm our expectations, thus illustrating the difficulty of hypothesizing interactions at the receptor level.

The heart was the experimental model chosen for the study of the relationship between tissue concentration and effect; it was seen to be sensitive to the adriamycin „dose-dependent" activity characterized by negative inotropic effect and decrease of both aortic and coronary flows. Among the various drugs (propanolol, quinidine, 8 β-carbobenziloxy-aminomethyl-1.6-dimethyl-10-ergoline maleate [MCE]) used to antagonize this toxic picture, only strophanthin, administered in the usual therapeutic doses, was able to prevent the occurrence of cardiotoxicity due to adriamycin or daunomycin when treatment preceded that with antitumoral drugs. This antagonism however depends on the dose of the two drugs.

Nevertheless, „*in vivo*", when the cardiotoxic picture due to adriamycin or daunomycin is already evident, strophanthin seems to allow a more rapid return to normal than that observed in animals treated with the antiblastic drug alone.

The protection exerted by strophanthin on the heart raises the problem of receptor-drug interactions in the cardiac district and naturally in the tumoral one too, suggesting the need to analyze possible variations in antineoplastic activity. Our experiments have excluded phenomena of antagonism on the action receptors at the neoplastic level and have revealed that the action receptors which interact with adriamycin in the tumor and the heart are influenced by digitalis glycoside in different ways. This experimental model has demonstrated that a knowledge of pharmacokinetics, combined with the study of specific pharmacodynamic effects, allows a safer use of a drug like adriamycin with a low chemotherapeutic index. A study of cardiotoxicity associated with antineoplastic therapy showed that the antitumoral effect of adriamycin, as well as that of daunomycin, is unaffected by administering the daily amount in two doses, while the dose-dependent cardiac effects are correspondingly lowered. The tendency of the antitumoral drug to accumulate in the heart could in a long run produce toxic effects, but the use of subdivided doses or pre-treatment with strophanthin, or both, should further decrease this possibility.

Summary

A study carried out by spectrofluorimetric techniques of pharmacokinetics, and an analysis of the pharmacodynamic properties of adriamycin in comparison with those of daunomycin, revealed side effects of the two drugs on the heart. In experiments performed on dogs, a rapid decrease of inotropism, coronary flow and aortic flow was observed following the administration of 1.5 mg/kg of the two antiblastic drugs, thus showing a dose-effect relationship. Strophanthin was able to antagonize the negative inotropic actions of the two antitumor drugs in the dog and in isolated rabbit hearts. The observed antagonism between the cardiac glycoside and adriamycin is believed to be due to reduced binding between the antiblastic drug and tissues.

The biochemical and biophysical tests carried out (determination of catecholamine content in the heart and suprarenal capsules, O_2 consumption by liver and heart homogenates, oxidative phosphorylation, CPK, Na^+K^+-activated, Mg^{++}-dependent ATPase activity, actomyosin superprecipitation, ^{45}Ca uptake and release by mitochondria and sarcotubules) did not enable us to detect the mechanism responsible for the toxic action of the two drugs on the heart.

References

1. ARENA, E., DUSONCHET, L., GEBBIA, N., GERBASI, F., PICONE, M. A., TRAINA, A.: Progress in antimicrobial and anticancer chemotherapy. (Procedings of the 6th International Congress of Chemotherapy) **2**, 124 (1969).
2. KOHN, K. W.: Anal. Chem. **33**, 862 (1961).
3. TUBARO, E., BULGINI, M., J.: Nature **212**, 1314 (1966).
4. DUSONCHET L., GEBBIA, N., GERBASI, F.: Pharmacol. Res. Comm. **3**, 55 (1971).
5. PASQUALINO, A., PICONE, M. A., TRAINA, A.: Arzneim. Forsch. (Drug Res.) **19**, 774 (1969).
6. BOLLUM, F. J.: J. Biol. Chem. **234**, 2733 (1959).
7. MANS, R. J., NOVELLI, G. D.: Arch. Biochem. Biophys. **94**, 18 (1961).
8. GIOTTI, A., JNGIANNA, J.: Arch. Ist. Sc. Farmacol. **9**, 502 (1959).
9. UMBREIT, W. W., BURRIS, R. H., STAUFFER, J. F.: Manometric Techniques. Burgess Publishing Co. Minneapolis (1957).
10. HUNTER, E.: Methods in enzimology. **2**, 610 (1955).
11. MATSUI, H., SCHWARTZ, A.: Biochim. Biophys. Acta. **128**, 380 (1966).
12. FISKE, C. H., SUBBAROW, Y.: J. Biol. Chem. **66**, 375 (1965).
13. EBASHI, S.: The J. of Biochem. **50** (3), 236 (1961).
14. FANBURG, B., GERGELY, J.: The J. of Biol. Chem. **240**, (6), 2771 (1964).
15. CASAZZA, A. M.: Corso superiore Chemiolerapia Tumori 337, Milano (1970).
16. DIMARCO, A., GAETANI, M., SCARPINATO, B.: Cancer Chemother., Rep. **53**, 33 (1969).
17. SILVESTRINI, R., GAMBARUCCI, C., DASDIA, T.: Tumori **56**, 137 (1970).
18. WANG, J., CHERVINSKY, D. S., ROSEU, J.: Proc. Amer. Ass. Cancer Res. **12**, (1971).
19. ARENA, E., D'ALESSANDRO, N., DUSONCHET, L., GEBBIA, N., GERBASI, F., PALAZZOADRIANO, M., RAUSA, L., TUBARO, E.: Arzneim. Forsch. (Drug Res.) **8**, 1258 (1971).

Pharmacokinetics and Metabolism of Adriamycin and Daunomycin

D. W. YESAIR, M. A. ASBELL, R. BRUNI, F. J. BULLOCK, and E. SCHWARTZBACH

Arthur D. Little, Inc., Cambridge, MA/USA

With 7 Figures

We have determined the physiological disposition of adriamycin and daunomycin in several animal species. Both drugs were rapidly cleared from plasma, deposited in tissues and then excreted slowly. Daunomycin was effectively metabolized *in vivo* by mice, rats, dogs, and hamsters, whereas no evidence was found for the metabolism of adriamycin except in hamsters. In the first few figures, I wish to compare the pharmacokinetics of daunomycin and adriamycin in the rat.

The concentration of daunomycin and adriamycin equivalents in plasma of rats decreased rapidly and remained nearly constant at a low level after 20 min. (Fig. 1) Similar data for daunomycin have been obtained for mice and dogs. These data indicate a localization of drug, perhaps, in tissues and/or excretory products.

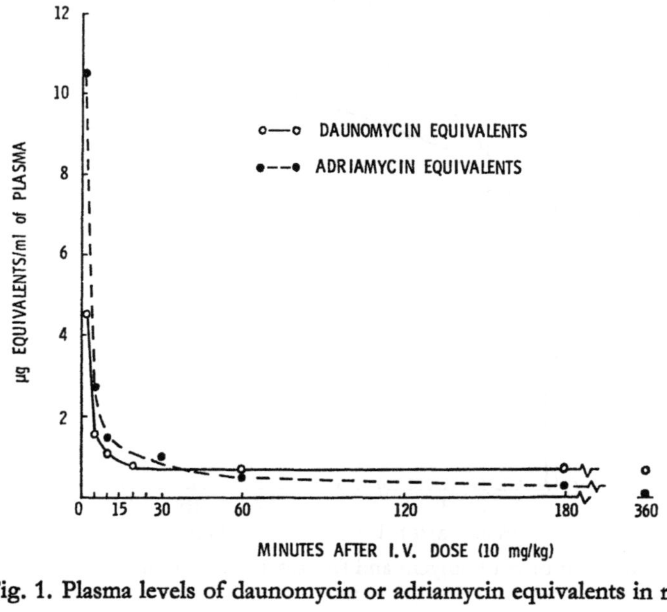

Fig. 1. Plasma levels of daunomycin or adriamycin equivalents in rats

The cumulative excretion of both daunomycin and adriamycin in urine by rats with cannulated bile ducts can be seen in Fig. 2. The excretion of daunomycin equivalents in urine and bile was essentially the same. The secretion of adriamycin into bile was slightly greater than its excretion into urine. In comparison to daunomycin, the excretion of adriamycin equivalents via the two routes was about one-third that of daunomycin equivalents in both normal and bile-duct cannulated rats.

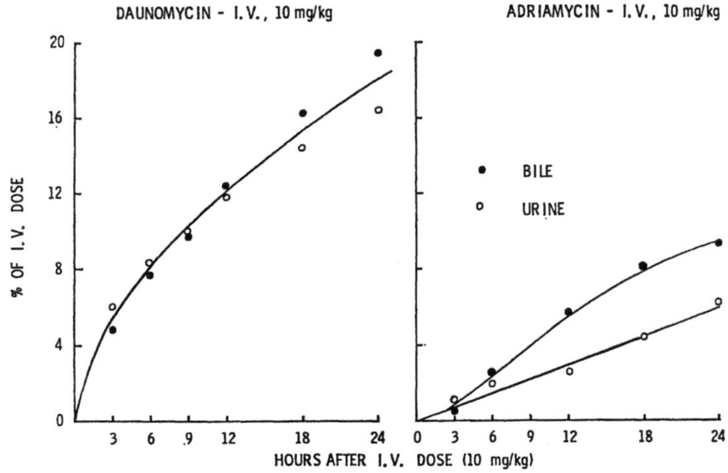

Fig. 2. Excretion of daunomycin or adriamycin by bileduct-cannulated rats

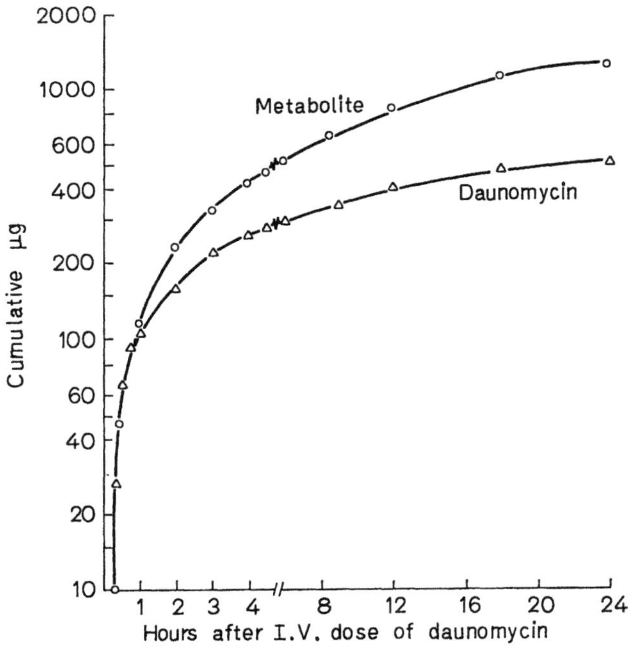

Fig. 3. Biliary excretion of daunomycin and fluorescent metabolite by bile duct-cannulated rats

The fluorescent drug equivalents in urine and in bile from rats have been fractionated by thin-layer chromatography and quantified by fluorescence spectrophotometry. Only two fluorescent species were present in urine and bile; i.e., daunomycin and a major metabolite. This major metabolite appears to correspond to daunorubicinol, D^2, recently described by BACHUR (1). The quantitative rate of excretion of Daunomycin and of this major metabolite *(D^2)* is represented here (Fig. 3). Daunomycin excretion into bile was initially greater than that of the metabolite. At 2h, the metabolite represented more than 50% of daunomycin equivalents. By 24 h, this metabolite was about two-thirds of the total excreted drug equivalents.

Studies on the adriamycin equivalents from urine of rats and mice showed no evidence of adriamycin metabolism; only a single fluorescent compound was found and its chromatographic behavior and fluorescence were similar to infused drug.

Tissues were analyzed in order to determine the general distribution of daunomycin and adriamycin as well as to determine the extent of metabolism of these drugs. The distribution of adriamycin, daunomycin, and metabolite, D_2, in several tissues of rats is shown in Fig. 4. Initially, the concentration of daunomycin was greater than

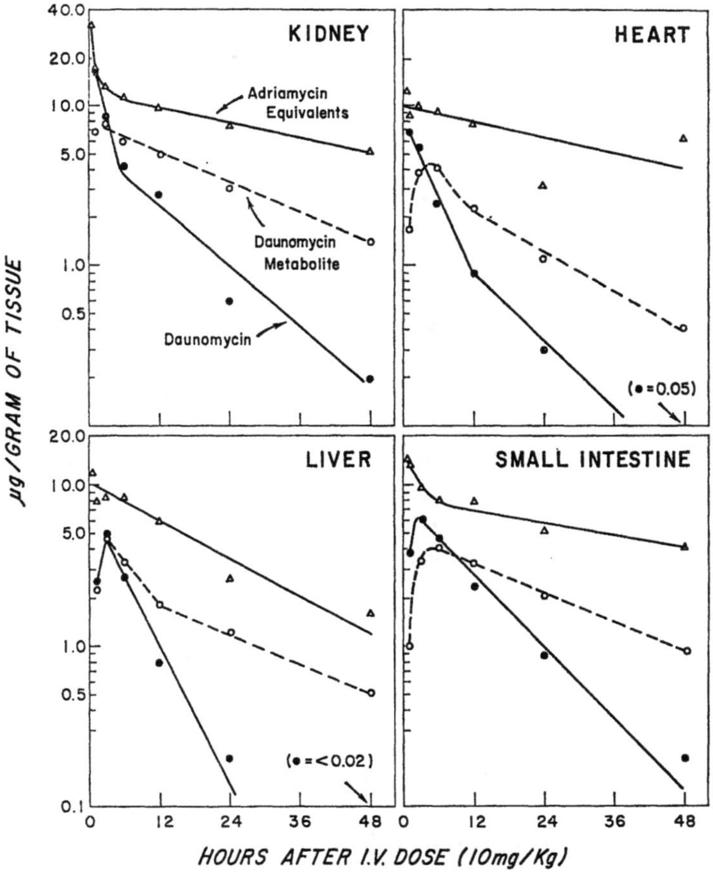

Fig. 4. Concentration of drug species in tissues from rats

metabolite but the concentration of metabolite became greater than that of daunomycin within 3—6 h. In all the tissues examined the concentrations of adriamycin equivalents were greater than the concentrations of daunomycin and its metabolites. The efflux of adriamycin equivalents appeared more similar to that of metabolite than to that of daunomycin. Tissue distributions of daunomycin and its metabolite in mice and dogs were similar to those in rats. No metabolite of adriamycin was detected in the tissues of any of these animals.

The calculated Cxt — that is, area under the concentration-time curve — of adriamycin equivalents in all four tissues of rats was several times that for daunomycin, its metabolite and their combined total. Kidney has the highest Cxt of daunomycin, its metabolite and adriamycin equivalents. The longer retention of adriamycin equivalents may correlate with the increased therapeutic index and with the findings that intermittent therapy is as effective as continuous daily treatment with adriamycin.

Table1. Calculated Cxt of daunomycin, its metabolite and adriamycin equivalents in tissues of rats

Tissue	Total Cxt (μg h/g) Daunomycin	Metabolite	Total dauno/met.	Adriamycin equivalents
Liver	38	87	125	199
Kidney	108	174	282	404
Heart	49	84	133	289
Small intestine	84	100	184	299

It should be emphasized that we have found only a single metabolite of daunomycin and no metabolite of adriamycin in tissues of normal rats, mice, and dogs. However, it has been reported by Bachur et al. (2) and Di Marco et al. (3) and confirmed by us, that tissue homogenates from rats converted daunomycin to other metabolites whose characteristics resembled the aglycones of daunomycin and its metabolite, D_2. In our studies *in vitro* we found two additional aglycones, D_x and D_y, which are chromatographically different from the aglycones of daunomycin and its metabolite, D_2. In addition, we found that cell-free liver homogenates of hamsters converted adriamycin to aglycone-like compounds. Thus far, we have found no adriamycin metabolite which is analogous to the D_2 metabolite. In the next few figures, I will briefly describe *in vitro* studies on the metabolism of daunomycin and adriamycin.

Fig. 5 depicts the rate of the aerobic conversion of daunomycin to D_2 by a cell-free system containing NADPH and the $100000 \times g$ supernatant of kidneys from rats. Within 5 min, 50% of the daunomycin had been transformed, and by the end of 30 min, the reaction had gone essentially (> 98%) to completion. Again, I wish to stress that this reaction occurs with the $100000 \times g$ supernatant under aerobic conditions.

Using a cell-free homogenate of kidney, NADPH and daunomycin under anaerobic conditions, we found that daunomycin appeared to be metabolized first to D_2 which, in turn, was metabolized to the unknown, designated D_x (Fig. 6). The

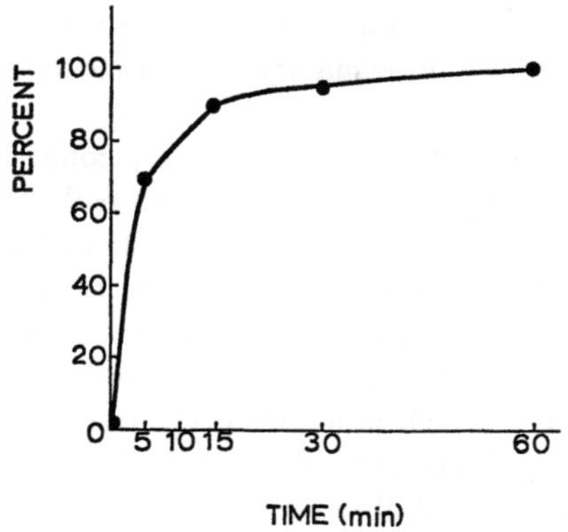

Fig. 5. Percent of metabolite (D_2) produced from daunomycin (D_1) by a soluble cell fraction of rat kidney incubated under aerobic conditions

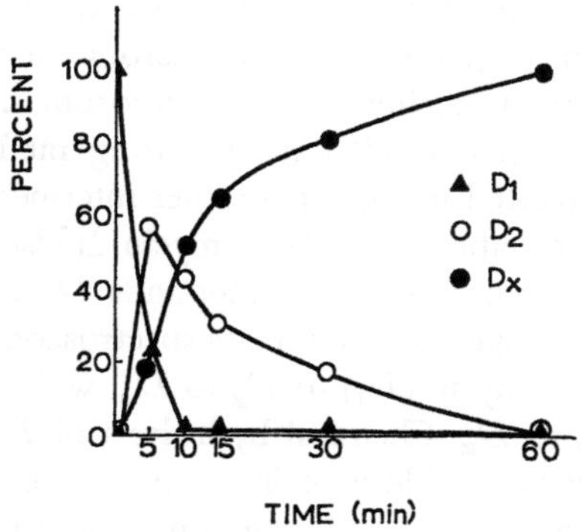

Fig. 6. Anerobic conversion of daunomycin (D_1) to the metabolite (D_x) with D_2 as a probable intermediate

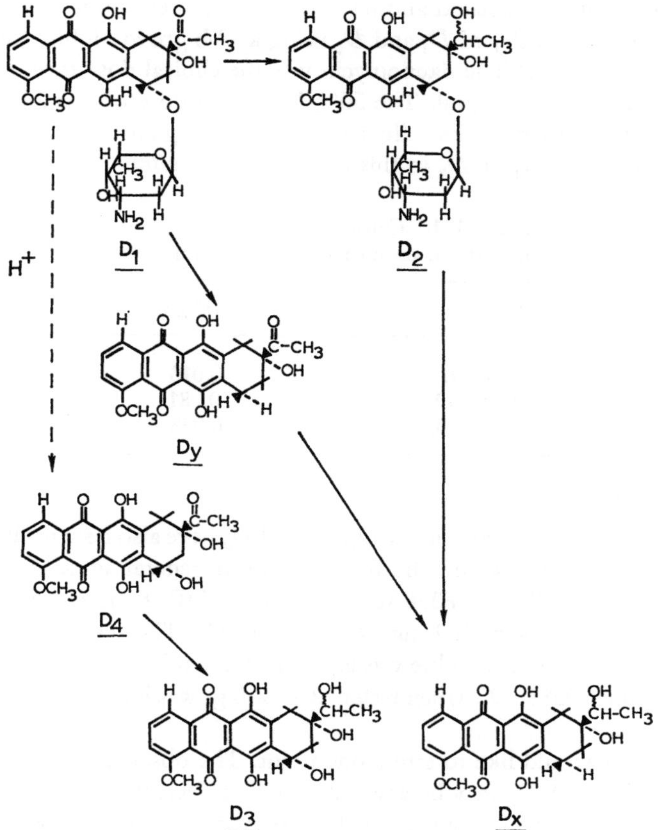

Fig. 7. Pathways for daunomycin metabolism by cell homogenates from rat livers and kidneys

anaerobic metabolism of isolated D_2 to D_x by all fractions was equivalent to that using D_1. Most of this anaerobic metabolism of daunomycin and D_2 is carried out by the mitochondria and microsomes.

Similar *in vitro* studies using rat liver homogenates showed that D_1 could be converted to D_x via another intermediate, D_y. Fig. 7 shows the in vitro pathways for metabolism of daunomycin. It also shows the chemical structures which we have determined for the compounds D_x and D_y. Structures are based on data from infrared spectroscopy, mass spectroscopy and NMR. Rat kidney homogenates convert daunomycin (D_1) to D_2 to D_x, while rat liver extracts carry out the reaction D_1 to D_y to D_x. Chemical hydrolysis of D_1 gives D_4, which is metabolized to D_3 by a kidney soluble protein fraction. Isolated D_3 can be converted to D_x by a liver or kidney microsomal fraction. Isolated D_y can be metabolized to D_x; however, the reverse reaction does not occur. Our preliminary studies with adriamycin and with homogenates of liver from hamsters indicate that adriamycin is converted to A_y which, in turn, goes to A_x. A_y and A_x are analogous to their daunomycin counterparts. Our cell-free studies indicated that the extent of conversion of D_1 both to D_2 and D_x was such that sufficient quantities of these two components could be produced for testing of their chemotherapeutic activity. The two compounds were enzymatically produced, isolated and tested for chemotherapeutic activity.

The relative cytotoxicity of D_1 and D_2 using a 48-h growth inhibition assay with L1210 is seen in Table 2. The daunomycin was the clinical dose form and the 50% inhibition dose was 0.013 μg/ml. The 50% inhibition dose of metabolite, D_2, was 0.081 μg/ml, or about 6 times less effective than the D_1 against L1210. D_x was much less active than either D_1 or D_2 in this assay system *in vitro*.

Table 2. L1210 Cytotoxicity Assay of Daunomycin and Metabolites (48-h growth inhibition)

Compound	ID_{50} (μg/ml)
Daunomycin	0.013
Metabolite D_2	0.081
Metabolite D_x	0.933

The chemotherapeutic activities of D_1, D_2 and D_x have also been evaluated against P388 cells growing *in vivo* and the results were in general agreement with those obtained *in vitro* with L1210. P388 tumor-bearing BDF_1 female mice were treated once a day for 10 days with 1 mg/kg of D_1 or D_2. The results seen in Table 3 demonstrate that the D_2 metabolite was equivalent in antileukemic activity to daunomycin (D_1) against P388. D_x, when tested at 1 mg/kg, was inactive against P388 cells *in vivo*.

At this time I would like to stress one point. The observed decrease in chemotherapeutic activity of D_x may be associated with a decreased uptake of D_x by the tumor. However, we could expect that this metabolite would be produced *in situ*. and consequently, may have potent chemotherapeutic activity as well as toxicological activity.

Table 3. Comparative effectiveness of D_1 and D_2 against P388 lymphocytic leukemia

Drug	Dose mg/kg	Δ, change body weight day 5	Median survival time (days)	% increase in lifespan
Daunomycin (D_1)	0	—	11.0	—
	1.00	−1.5	18.0	63
	1.25	−2.2	16.0	45
Metabolite (D_2)	0	—	10.0	—
	1.00	−3.6	15.5	55

With regard to production of these metabolites *in situ*, MHATRE et al. (4,5) found aglycone-like compounds in tissues of hamsters dosed with daunomycin and adriamycin and noted that the chemically prepared aglycones of daunomycin and adriamycin produced cardiac toxicity in hamsters and dogs. We have undertaken a preliminary study of the physiological disposition of daunomycin and adriamycin in hamsters and have observed that aglycone-like compounds of adriamycin and daunomycin as well as the normal metabolite, D_2, of daunomycin are present in liver. Based upon our chromatographic data, we conclude that the aglycone-like material in tissues from daunomycin-treated hamsters corresponded in R_F to the aglycone, D_x, which was found to be produced by tissue homogenates under anaerobic conditions. Chromatographically, the aglycone-like compound found in tissues from adriamycin-treated animals is similar to the A_y in our *in vitro* studies.

The significance of the differences in the metabolism of daunomycin and adriamycin may have relevance to both the chemotherapeutic and toxicological differences seen with the two compounds.

Acknowledgements

This work was supported by Contract PH 43-65-61 with Chemotherapy, National Cancer Institute, NIH. Daunomycin was provided by Drug Research and Development, NCI. Adriamycin was kindly supplied by Farmitalia, Milano, Italy.

References

1. BACHUR, N. R.: J. Pharmacol. exp. Ther. **177**, 573 (1971).
2. BACHUR, N. R., CRADOCK, J. C.: J. Pharmacol. exp. Ther. **175**, 331 (1970).
3. DIMARCO, A., BORETTI, G., RUSCONI, A.: Farmaco, Ed. sci. **22**, 535 (1967).
4. MHATRE, R. M., HERMAN, E., HUIDOBRO, A., WARAVDEKAR, V. S.: Pharmacologist **12**, 243 (1970).
5. MHATRE, R. M., HERMAN, E., HUIDOBRO, A., WARAVDEKAR, V. S.: J. Pharmacol. exp. Ther. **178**, 216 (1971).

Blood Levels, Tissue Distribution and Clinical Effects of Adriamycin

KIYOJI KIMURA, HIROSHI FUJITA and YASUNOBU SAKAI

National Cancer Center Hospital, Tokyo, Japan

With 5 Figures

We have been making comparative studies of in vivo distribution and clinical effects of many anticancer drugs. In this paper we present our experimental results and some clinical effects of adriamycin with special reference to its in vivo distribution.

I. Experimental Results of Blood Levels and Tissue Distribution

Our results reported earlier showed that anticancer drugs with a similar mode of action had largely common patterns in their blood levels, tissue distribution, excretion and inactivation. Therefore, the body distribution of adriamycin was compared to that of other drugs with a similar mode of action such as daunomycin, acetyl-kidamycin, actinomycin D and chromomycin A_3. Acetyl-kidamycin is a new anticancer antibiotic, discovered by Prof. HATA in Japan.

These 5 drugs including adriamycin have fairly strong activities against Gram-positive bacteria. Therefore, for this study of their body distribution, antibacterial activity was measured by microassay, the horizontal diffusion method or cup method being employed on very thin agar plates inoculated with spores of Bacillus subtilis, ATCC 6633.

Furthermore, in several experiments, the tissue distribution of adriamycin was also investigated using ^{3}H-labeled adriamycin. The standard curve of adriamycin in the bioassay is shown in Fig. 1. Blood levels of adriamycin in rabbits were measured by this method. As shown in Fig. 2, the maximum blood level is very low, and it decreases rapidly and disappears a few hours after the administration of adriamycin, but a very small amount of adriamycin reappears several hours later. This rebound curve is characteristic. This phenomenon may be caused by reabsorption of adriamycin excreted into the bile through the intestine or by elution of adriamycin fixed in the various tissues.

Blood levels of actinomycin D and daunomycin in rabbits are also shown in Fig. 3. Low and rebound blood levels are commonly observed in these drugs. Their rebound curves appearing several hours later are somewhat higher than for adriamycin. The height of the rebound curve is assumed to correlate with the reabsorption of the drug excreted into the bile via the intestine. Blood levels of chromomycin A_3 and acetyl-kidamycin also show the same rebound patterns.

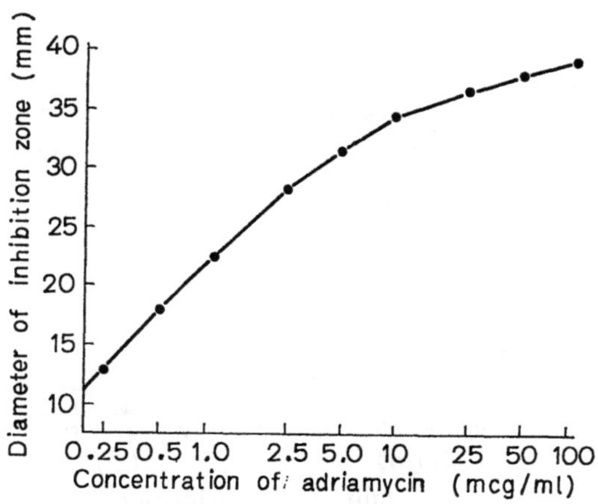

Fig. 1. Standard curve of adriamycin. Thin plate-cup method B. subtilis, ATCC6633

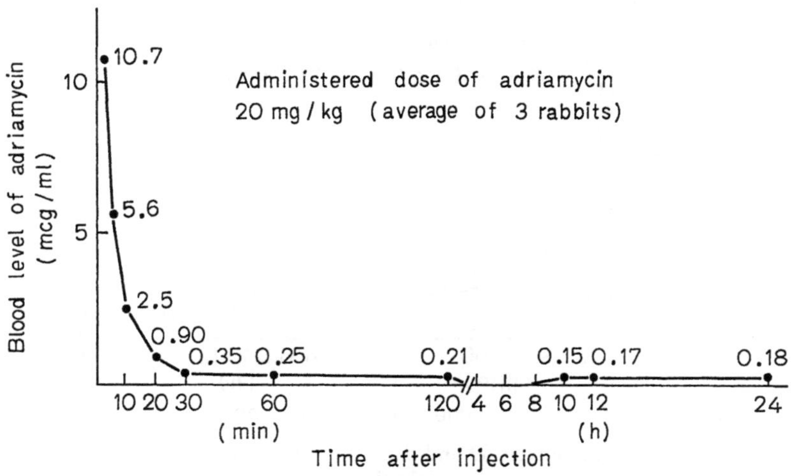

Fig. 2. Blood levels of adriamycin in normal rabbits after intravenous injection. Administered dose of adriamycin 20 mg/kg (Average of 3 rabbits)

If we arrange the drugs in order from the highest blood level at the same dose, chromomycin A_3 comes first, then actinomycin D, acetyl-kidamycin, daunomycin and adriamycin. These results are considered to be in inverse proportion to the adsorption rates of drugs by tissues. In other word, adriamycin shows the lowest blood level and the highest adsorption rate among these 5 drugs. The tissue levels of adriamycin in mice, measured by antibacterial activity, are shown in Table 1.

The level of adriamycin is high in tissues of the kidney, liver, intestine, lung, spleen, stomach, thymus and heart in both normal and tumor-bearing mice, and is maintained in these tissues 5 h after the administration of adriamycin. Futherore, as Table 1 shows, a large amount is excreted into the bile.

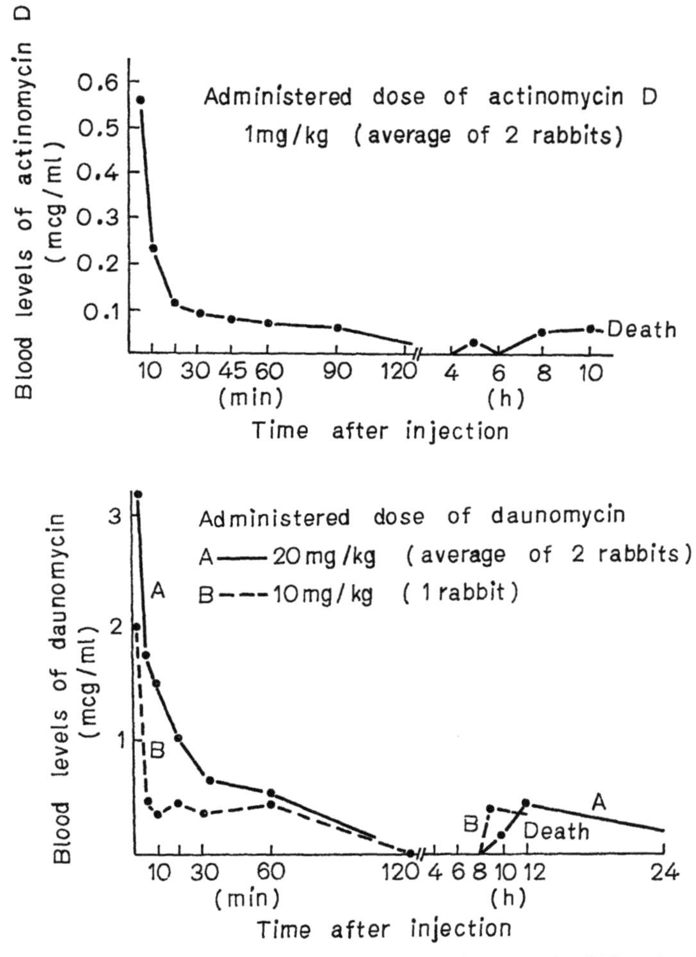

Fig. 3. Blood levels of actinomycin D and daunomycin in normal rabbits after intravenous injection

Tissue levels of adriamycin, daunomycin, and acetyl-kidamycin in sarcoma 180-bearing mice are shown in Table 2. Adriamycin and daunomycin show a common pattern in tissue level and high excretion into bile. At the same dose administered, the highest tissue levels in various organs were observed with acetyl-kidamycin, followed by daunomycin, and then adriamycin. This is assumed to relate the adsorption rate of each drug by the tissues. Drugs highly adsorbed did not show the large inhibition zone in bioassay. Drug levels in tumor tissue of sarcoma 180 are not as high as in kidney, liver, lung and spleen.

Table 1. Tissue levels[a] of adriamycin in mice after intravenous injection. Mice: ddN♂ (A — normal mice, B — sarcoma 180 bearing mice.) Administered dose of adriamycin: 50 mg/kg

Time after injection (h)	A 0.5	1	2	5	B 2
Kidney	6.1	7.1	5.3	5.5	4.7
Liver	2.8	3.0	3.3	2.8	3.6
Intestine	2.9	2.8	2.3	2.5	3.1
Lung	2.5	1.9	2.3	2.4	2.7
Spleen	2.5	1.9	2.5	2.8	1.1
Stomach	—	1.8	0.7	3.9	1.5
Thymus	—	0.75	2.0	—	1.2
Heart	0.9	0.7	0.8	—	1.0
Pancreas	—	—	0.98	—	4.1
Testis	—	—	—	—	—
Muscle	—	—	—	—	—
Skin	—	—	—	—	—
Blood sediment	—	—	—	—	—
Serum	1.6	0.80	0.45	0.40	0.37
Urine	27.7	60.0	19.4	13.9	16.0
Bile	67.5	114.0	109.0	98.0	71.0
Tumor (Solid)					1.2

[a] mcg/g or mcg/ml

Table 2. Tissue levels[a] of 3 anticancer drugs in sarcoma 180-bearing mice. Administration method: I.V. injection, dose: 50 mg/kg. Time: 2 h after administration

	Adriamycin	Daunomycin	Acetyl-kidamycin
Heart	1.0	3.7	—
Lung	2.7	3.3	24.0
Thymus	1.2	—	4.0
Liver	3.6	4.0	3.8
Spleen	1.1	3.4	7.9
Kidney	4.7	5.3	7.6
Pancreas	4.1	—	7.5
Stomach	1.5	2.3	—
Intestine	3.1	4.0	7.4
Testis	—	—	—
Muscle	—	—	—
Skin	—	—	—
Blood sediment	—	3.7	—
S. 180 (Solid)	1.2	1.2	2.2
Serum	0.37	0.79	1.7
Urine	16.0	14.5	7.7
Bile	71.0	99.0	15.0

[a] mcg/g or mcg/ml.

It is possible that the high concentrations of the drugs in the bile are reabsorbed through the intestine and indicate so-called enterohepatic circulation. But this circulation is not considered to be apparent in adriamycin, because the rebound level of adriamycin is very low. Moderately high concentrations of these drugs are excreted in the urine. The distribution of ^{3}H-labeled adriamycin among organs of sarcoma 180-bearing mice are shown in the Table 3. Drug level was measured by both radioactivity counting (A) shown in Table 4 and antibacterial methods (B). The radioactivity per gram of various organs in descending order was: kidney, liver, lung,

Table 3. Tissue levels of adriamycin in mice. Distribution of ^{3}H-adriamycin among organs of 2 mice. 1 h after intravenous injection of 50 mg/kg in the sarcoma 180-bearing mice

Organs	Radioactivity[a] dpm/g (x 10^{3})	Radioactivity[a] mcg/g (A)	Antibacterial activity[b] mcg/g (B)	B/A
Kidney	8057	395.4	3.3	0.008
Liver	6489	318.4	2.6	0.008
Lung	5679	279.6	1.9	0.006
Heart	4066	199.6	—	
Spleen	3597	176.5	1.6	0.009
Intestine	2608	128.0	1.6	0.012
Stomach	2208	108.4	1.7	0.015
Thymus	1397	68.5	1.6	0.023
Muscle	1086	53.3	—	
Skin	954	46.8	1.1	0.023
Testis	336	16.5	—	
Serum	238	11.7	—	
Bile				
Ascites	101	4.9	—	
Tumor cell	392	20.9	—	

[a] Radioactivitiy of ^{3}H-ADM: 20378 dpm/mcg.
[b] Test organism: B. subtilis, ATCC 6633.

Table 4. Materials and methods for in vivo distribution studies of ^{3}H-adriamycin

1 Animal:	Sarcoma 180-bearing ddN mice Number of mice 4 weight 19.0-27.5 g
2 Specific activity of ^{3}H-Adriamycin:	23.2 μci/mg
3 Dose:	50 mg/kg, 1.16 mci/kg
Route of injection:	Vena caudalis
4 Samples:	1 h and 5 h after injection of ^{3}H-Adriamycin, 2 mice were killed and 10-200 mg samples were taken from various tissues at each time
5 Radioactivity counting:	After oxidation with ^{3}H oxidizer (Packard model 300), radioactivities of samples were counted by liquid scincillation spectrometer (Packard model 3320)
6 Tissue autoradiography:	Fixation; formalin solution Dipping; Sakura Emulsion NR-M 2 Exposure; 2 weeks and 4 weeks

Table 5. Distribution of ^{3}H-adriamycin among organs of 2 mice. 5 h after intravenous injection of 50 mg/kg of the drug in sarcoma 180-bearing mice

Organs	Radioactivity[a] dpm/g (x 10³)	Radioactivity[a] mcg/g (A)	Antibacterial activity[b] mcg/g (B)	B/A
Kidney	6327	310.5	4.7	0.015
Liver	6488	318.4	2.0	0.006
Lung	5301	260.1	1.6	0.006
Heart	2745	134.7	—	
Spleen	4208	206.5	2.5	0.012
Intestine	2144	105.2	2.2	0.021
Stomach	3010	148.8	1.5	0.010
Thymus	2332	114.4		
Muscle	1177	57.7	—	
Skin	897	44.0	—	
Testis	665	32.6	—	
Serum	124	6.1	—	
Bile	13653	670.0		
Ascites	45	2.2	—	
Tumor cell	474	23.3	—	

[a] Radioactivity of ^{3}H-ADM: 20378 dpm/mcg.
[b] Test organisms: B. subtilis, ATCC 6633.

Table 6. Adsorption of 3 anticancer drugs by mouse tissue homogenates

	Rate of loss of antibiotic activity (%)		
	Adriamycin	Daunomycin	Acetyl-kidamycin
Thymus	99.84	99.01	
Spleen	99.54	98.76	94.70
Pancreas	99.75	98.23	97.50
Testis	99.72	98.66	94.70
Liver	99.52	98.35	95.80
Kidney	99.32	98.08	95.70
Lung	99.10	98.08	92.00
Stomach	99.10	97.40	97.88
Intestine	98.64	97.10	98.97
Heart	96.50	95.40	94.00
Skin	88.30	88.80	87.30
Muscle	82.80	85.00	89.80
Blood sediment	64.00	66.00	80.40
Serum	38.69	26.75	59.20
S. 180	99.61	98.94	97.00
Ehrlich C.	100	98.88	94.50

Final concentration of drugs — 50 mcg/ml, Tissue H — 10%

heart, spleen, intestine, stomach, thymus, skin, tumor cells, testis, serum and ascites. These findings are in fairly good agreement with those obtained by bioassay. The organs having little radioactivity, such as muscle, skin, testis, serum, tumor cells, and ascites, did not show any antibacterial activity. The ratio B/A is approximately 1/100. This means that about 99% of the adriamycin is adsorbed by these tissues,

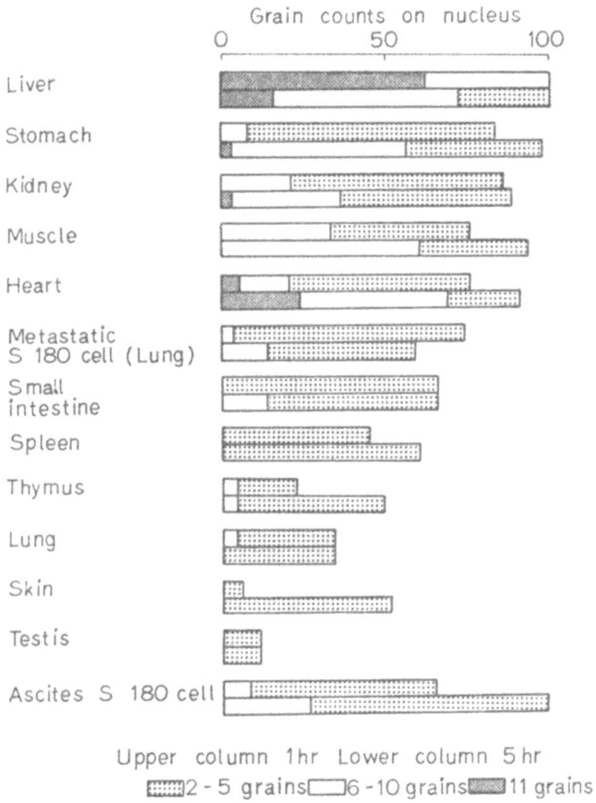

Fig. 4. ^{3}H-adriamycin distribution in tissue of sarcoma 180 bearing mouse. 1 h and 5 h after intravenous injection (50 mcg/g, 1.16/μci/g)

and adsorbed adriamycin shows a marked reduction in the antibacterial activity. The drug levels 5 h after the administration of ^{3}H-adriamycin are shown in Table 5. Organ levels showed no marked differences between 1 and 5 h. Such long-maintained tissue levels are characteristic of drugs that inhibit DNA-dependent RNA synthesis. This is thought to be due not only to high adsorption of the drugs by tissues but also to weak or slow inactivation by the enzyme. The adsorption rates of adriamycin, daunomycin and acetyl kidamycin by mouse tissue homogenate are shown in Table 6. These drugs are very rapidly and strongly adsorbed by the tumor tissues, thymus, spleen, pancreas, testis, liver, kidney, lung and digestive organs, but are not absorbed by serum. Adsorption of the drugs by these tissue homogenates proceeds even at

5 °C in a few minutes and reduces antibacterial activity. The highest rate of adsorption was observed in adriamycin, followed by daunomycin, and then acetyl-kidamycin. Rates for actinomycin D and chromomycin A_3 are lower. ^{3}H-adriamycin distribution in tissues of sarcoma 180-bearing mice was investigated using the autoradiographic technique as shown in Table 4. Grains are counted on nucleus and cytoplasm of liver, stomach, kidney, muscle and heart muscle cells etc.

Grain counts on nucleus are several times higher than that on cytoplasm in every cell of all of the tissues 1 h after the administration of ^{3}H-adriamycin. Similar results are observed in all cells except the liver cell even 5 h after the injection of ^{3}H-adriamycin. These results are shown in Fig. 4 and 5.

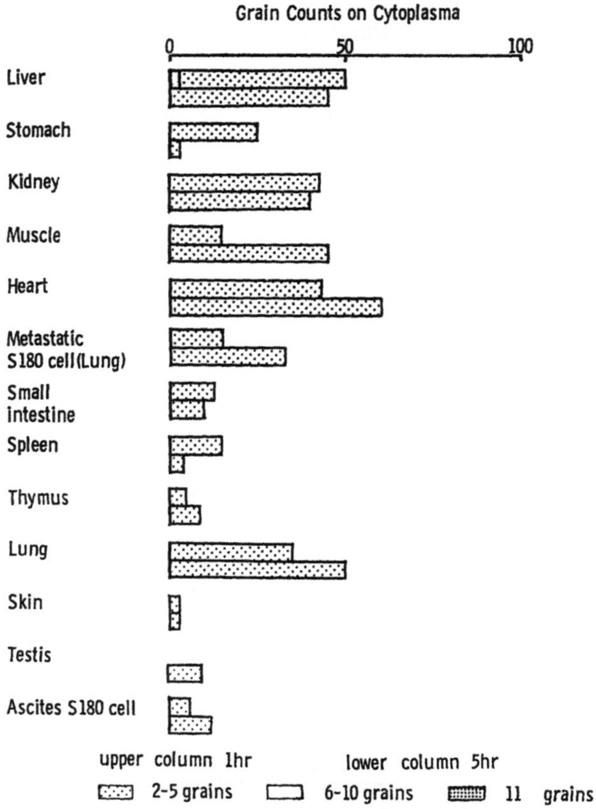

Fig. 5. ^{3}H-adriamycin distribution in tissue of sarcoma 180 bearing mouse. 1 h and 5 h after intravenous injection (50 mcg/g, 1.16/μci/g)

Furthermore, in vitro ^{3}H-adriamycin uptake in AH_{13} cells (Table 8) and human bone marrow cells (Table 9) are also investigated using, the single-cell autoradiographic technique as shown in Table 7. The results reveal that labeling index of AH_{13} tumor cells, leukemic cell and other cells is 100%.

Table 7. Materials and methods of in vitro single-cell autoradiography of ^3H-adriamycin

1 Materials:	a) Ascites hepatoma AH13 cells b) Human leukemic cells c) Normal bone marrow cells
2 Specific activity of adriamycin:	23.2 μci/mg
3 Incubation:	concentration; 1 μci and 2 μci/ml 43 μg (adriamycin)/ml and 86.2 μg/ml time; 5' and 30'
4 Autoradiographic technic:	Fixation; methanol Dipping; Sakura emulsion NR-M2 Exposure; 5 weeks

Table 8. In vitro ^3H-adriamycin uptake in AH13 ascites tumor cell

	5 min incubation		30 min incubation	
	Labeling index (%)	Median grain	Labeling index (%)	Median grain
1μci/ml	100	95	100	84
2μci/ml	100	80	100	75

Table 9. In vitro ^3H-adriamycin uptake in human bone marrows cells

			Labeling index %	Median grain
Case 1	AML	Leukemic cell	100%	27
		Lymphocyte	100%	17
Case 2	Anaplastic Carcinoma	Myelocyte	100%	54
		Metamyelocyte	100%	41
		Stab & Segmented	100%	37
		Lymphocyte	100%	34

1 μci/ml, 30 min incubation

II. Clinical Results

At the National Cancer Center Hospital, Tokyo, from January 1971 to July 1971, 19 cases with inoperable and advanced cancer were treated with adriamycin. These comprised: stomach cancer 8 cases; anaplastic carcinoma 2 cases; malignant lymphoma 4 cases, and others. Primary site of tumor, prior chemotherapy, daily doses and total doses, tumor effects and side effects are summarized in Table 10. Tumor effects were observed in 2 out of 8 cases with stomach cancer, 2 out of 2 cases of anaplastic carcinoma, 2 out of 4 cases of malignant lymphoma, and 2 out of 5 cases with other cancer sites, as shown in the Table 11. Loss of hair, diarrhea, stomatitis and nausea as side effects were each observed in 3 out of the 19 cases, as shown in

Table 10. Clinical results of cases treated with adriamycin

Pt. No.	Tumor type	Prior chemotherapy*	Daily doses	Total doses	Tumor effects	Side effects
1	Stomach	MMC, EX, FAMT, FAMC	20 mg × 3	60 mg	None	Loss of hair, diarrhea
2	Stomach	MTX	20 mg × 2, 20 mg × 2, 20 mg × 2	120 mg	None	
3	Stomach	MMC, FQMT, Pachyman	5 mg × 6, 10 mg × 4	70 mg	Slight	Stomatitis, diarrhea
4	Stomach		10 mg × 4, 10 mg × 3	70 mg	None	Arrhythmia
5	Stomach	FAMT, FQMT	40 mg × 4/w	160 mg	None	
6	Stomach		10 mg × 4, 10 mg × 4	80 mg	None	
7	Stomach		10 mg × 4, 10 mg × 2	60 mg	Good	Arrhytmia
8	Stomach		10 mg × 4, 10 mg × 3, 10 mg × 3	100 mg	None	
9	Hepatoma		20 mg × 2, 20 mg × 1, 10 mg × 4	100 mg	None	
10	Bile duct	VCR	20 mg × 1, 20 mg × 1	40 mg	None	
11	Rectum	MMC, FAMT	10 mg/2d × 3	30 mg	Slight	
12	Unknown (anaplastic ca.)	VEMP	20 mg/2d × 5, 20 mg × 1	120 mg	Marked	Loss of hair, tachycardia
13	Lung (anaplastic ca.)		10 mg × 7	70 mg	Moderate	
14	Breast		20 mg/2d × 3	60 mg	None	Stomatitits, nausea
15	Hemangio-pericytoma	VEP, V*EP, FQMT, BEP	20 mg × 2, 20 mg × 2, 20 mg × 2	120 mg	Slight	Nausea, palpitation
16	RCSA	VEMP, V*EMP, BLM, DCM, DCMP, etc.	40 mg/w × 2, 20 mg/w × 1	100 mg	None	Herpes zoster
17	RCSA	BLM, VEMP, V*EP, VENP, L-asp., etc.	20 mg × 2	40 mg	None	Nausea, diarrhea
18	RCSA	V*EMP, BEMP, VCR, VEMP	40 mg × 3	120 mg	Marked	
19	LSA	VEMP, BLM, DCMP, etc.	20 mg/2d × 3	60 mg	Marked	Stomatitis, loss of hair

* Abbreviations: MMC; Mitomycin C, EX; Endoxan, FAMT; 5-FU, EX, MMC, Toyomycin in combination, VCR; Vincristine, FAMC; 5-FU, EX, MMC, Cytosine arabinoside in combination, FQMT; 5-FU, Carbazil-quinone, MMC, Toyomycin in combination, VEMP; VCR, EX, 6MP, Prednisolone in combination, V*EP; Vinblastine, EX, Prednisolone, in combination, BEP; Bleomycin, EX Prednisolone in combination, BLM; Bleomycin, DCMP; Daunomycin, Cytosine arabinoside, 6MP, Prednisolone in combination, VENP; VCR, EX, Natulan, Prednisolone in combination.

Table 11. Tumor effects of adriamycin

Primary sites	Tumor effect			Rates
	Marked	Moderate	Slight	
Stomach		1	1	2/8
Anaplastic Ca.	2			2/2
Malignant lymphoma	2	0	0	2/4
Others			2	2/5
Total	4	1	3	8/19 (47.3%)

Table 12. Side effects in 19 cases treated with adriamycin

	No. of cases
Loss of hair	3
Diarrhea	3
Stomatitis	3
Nausea	3
Arrhythmia	2
Tachycardia	1
Papitation	1
Herpes zoster	1

Table 12. Arrhythmias were also observed in 2 of the 19 cases. Daily doses ranged from 5—40 mg/day; tumor effects were observed in the cases of carcinoma of stomach, lung, and rectum.

From these clinical results and other experimental results not presented in this paper, tumor effects should be expected even in cases treated with small daily doses of adriamycin.

The chemotherapeutic effects of adriamycin will be enhanced and the side effects reduced when more appropriate schedules of daily and total doses are found.

We believe that adriamycin is one of the most effective anticancer antibiotics.

Part V

Clinical Activity and Side Effects

Adriamycin in the Treatment of Acute Leukemia

H. Begemann and G. Wernekke

I. Medizinischen Abteilung des Städt. Krankenhauses München-Schwabing

With 1 Figure

In November 1969 we began to use adriamycin in the treatment of acute leukemia.

We randomized our patients (aged 15 to 65) (Table 1) into three groups, and each group was treated with a single drug regimen (adriamycin, Daunoblastine, cytosine arabinoside). Most of the patients had been pretreated with other antiblastic drugs.

Table 1

Age of the patients	< 20 years			20–30 years			30–40 years			40–50 years			> 50 years		
Results	*T*	*R*	*F*	*T*	*R*	*F*	*T*	*R*	*F*	*T*	*R*	*F*	*T*	*R*	*F*
	5			8			8			2			3		
aML		3			4	3		5	2			1			1
aLL			2												
cML					1				1						2
Erythrol.												1			

T = total number of treatments,
R = number of remissions,
F = treatment failures.

The dose of 0.5 mg/kg/d of adriamycin was administered for three consecutive days and the course repeated after an interval of seven days. Daunoblastine was given at a dose of 1.5 mg/kg on days 1, 8 and 16. Cytosine arabinoside was administered at the dose of 3 mg/kg/d for 5 consecutive days followed by a 10-day rest interval (Fig. 1).

The patients were also given 1.5 mg/kg/d of prednisone for the first 14 days, followed by a slowly reduced dosage. No patient was given more than 3 courses, but when a patient did not respond to a drug he was given another treatment. Careful physical examination, the peripheral and bone marrow picture, all the routine laboratory tests (transaminases, LDH, bilirubin, urea, creatinine, uric acid, fibrinogen, electrolytes, serum electrophoresis) were carried out before the start of treatment and repeated with various frequencies throughout it.

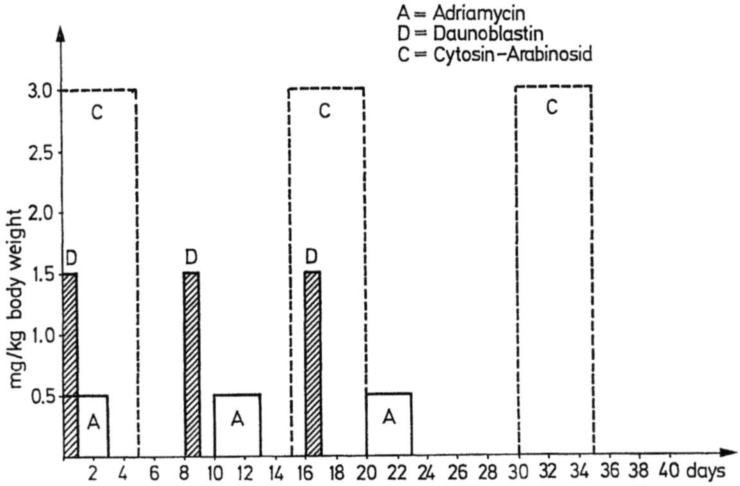

Fig. 1

31 courses of therapy were given to 28 patients, of whom 5 were inadequately treated because they died before the end of the second course. Eight out of 26 patients were adequately treated and had a complete remission, and 5/26 had a partial remission (as defined by the "Paul Ehrlich Gesellschaft" for chemotherapy) (Table 2). Adriamycin was used as remission inducer, but Endoxan was used for maintenance therapy in 2 patients, Puri-Nethol in 6, Methotrexate in 3 and Vincristine in 2.

Table 2. 26 adequate treatments

Remissions		Failure to therapy
Partial remissions	complete remissions	
5	8	13

We did not use any immunotherapy.

4 of the 5 patients with partial remission were suffering from acute myeloblastic leukemia, either in the initial phase or in second relapse. Remission lasted in one subject 152 days, in another 71 days and in two patients it still continues. One patient in the blastic phase of chronic myelocytic leukemia had a remission of 20 days only (Table 3). Partial and complete remissions are considered together in this study and the frequency of relapses and their distribution among the different forms of leukemia are reported in Table 4. It is stressed that a remission was achieved in 5 patients in their second relapse, i.e. 7—24 months after diagnosis, and in one case in the third relapse, 11 months after diagnosis. Two patients with acute lymphoblastic leukemia, in second and third relapse respectively, did not respond to adriamycin treatment.

No constant correlation was observed between the number of relapses before treatment and the induction and length of remission.

Table 3

Duration of remission (days)	Initial treatment	1. Relapse	2. Relapse	3. Relapse
cML	20			
aML				24
		46	47	
			50	
			71	
			86	
	118			
	150			
			152	
	Continues 60 and 30			
		Continues 20		

Table 4

	Onset of disease			1. Relapse			2. Relapse			3. Relapse		
Results	*T*	*R*	*F*	*T*	*R*	*F*	*T*	*R*	*F*	*T*	*R*	*F*
aML	8	4	4	2	2		5	5		4	1	3
aLL							1		1	1		1
cML	2	1	1	2		2						
Erythrol.	1		1									

T = total number of treatments,
R = number of remissions,
F = treatment failures.

Alopecia was the most frequent side effect observed (in males beard growth ceased) especially after the second week of therapy. Hair began to grow again from the fourth to sixth week after the end of the therapy.

Inflammation and ulceration of mucous membranes in mouth, throat and oesophagus were another frequent side effect which caused great discomfort to the patients, sometimes worsening the general condition by making eating difficult. This condition was usually present in the second week of treatment and often caused therapeutic problems because of the onset of secondary bacterial and mycotic infections related to the lowered resistance of these patients.

Special care together with supportive and prophylactic therapy had to be initiated to overcome this condition. Topical inflammation of the vein and topical irritation of tissues at the site of the injections were frequently observed. No cardiotoxicity reaction such as tachycardia, pulmonary oedema or electrocardiographic irregularities were seen, although ECG was done weekly.

No persistent changes in the biochemical data were noted; it is uncertain whether Ihe reversible increase in transaminases or the decrease in proteins, especially in tg, were to be ascribed to the disease or to the therapy.

A remarkable depression of blast cells in bone marrow and in peripheral blood was observed in all the cases, as well as of the other lines of blood cells. Such pancytopenia set in usually on the 14th to 18th day after the start of the treatment and was more pronounced in patients who came into remission; this was heralded by an increase in normal granulocytes and platelets in the peripheral blood. The bone marrow appeared normal 25—30 days after the beginning of therapy. It is of practical interest to determine whether there is a crossresistance beetwen daunomycin and adriamycin. This question cannot be answered from our data. We observed that out of 13 patients in remission after adriamycin therapy 4 patients had previously been treated successfully with daunoblastine alone and 3 others successfully with Daunoblastine in combination with L-asparaginase. One patient who had been unsuccessfully treated with Daunoblastine alone, however, responded to adriamycin. Another patient who had previously been successfully treated with Daunoblastin alone, as well as 2 patients who had been treated successfully with Daunoblastine in combination with asparaginase, did not respond to our treatment with adriamycin. Three patients who came into remission after adriamycin in the 2nd relapse failed to respond in the 3rd relapse.

In conclusion, our results in a randomized treatment of acute leukemia were the following: the remission rate with adriamycin was 50%, with Daunoblastin 25.8% and with cytosine-arabinoside 24.2% (Table 5).

Table 5

	Treatments	Adequate treatments	First treatment	Remissions	Remission after first treatment
Adriamycin	31	26 100%	11 42,3%	13 50 %	5
Daunoblastin	36	31 100%	11 35,4%	8 25,8%	4
Cytosine-arabinoside	38	33 100%	8 24,2%	8 24,2%	2

Clinical Trials with Adriamycin. Results of Three-Years Study

GIANNI BONADONNA, SILVIO MONFARDINI, MARIO DE LENA, FRANCA FOSSATI-BELLANI and GIANNI BERETTA

Istituto Nazionale per lo Studio e la Cura dei Tumori, Milano

With 4 Figures

The results of the initial clinical studies (Phase I and preliminary Phase II evaluation) carried out at the National Cancer Institute of Milan an adriamycin, a 14-hydroxy analogue of daunomycin [1, 17], have been reported since 1969 in numerous papers and meetings [2-7, 9-12, 24-26]. Since favorable therapeutic results appeared evident from the first series of patients, the trials have been actively pursued and the drug was tested on a large population of patients with different forms of neoplastic disease.

During the past year, when we had sufficient evidence of the effectiveness of the drug against certain types of neoplastic disease, adriamycin was introduced into various protocols of combination chemotherapy (acute lymphoblastic leukemia, acute lymphosarcoma cell leukemia, malignant lymphomas, solid tumors of children, soft tissue sarcomas of adults, carcinomas of testicle and carcinoma of lung). The aim of the present report is to summarize our clinical experience from August 1968 to June 1971 with the use of this new antitumor antibiotic, administered intravenously, alone and in combination with conventional drugs. The results of adriamycin administered by prolonged arterial infusion have been reported elsewhere [18] by other members of this Institute.

Material and Methods

Patient Selection

The first series of about 100 patients (Phase I evaluation) had far advanced non-resectable or extensively metastatic malignant neoplasms, although their performance status [19] was in no instance below 40. Most of them had already received, prior to the administration of adriamycin, one or more courses of radiotherapy and/or chemotherapy. The second series of patients (Phase II evaluation) had less extensive metastatic disease and only the minority of cases were pretreated with conventional agents. The third series includes patients given adriamycin in combination with other drugs.

They were usually untreated, and in most instances showed a limited extent of metastatic disease. Table 1 summarizes the whole case material treated with adriamycin during the past three years.

Table 1. Case material treated with adriamycin from August 1968 to June 1971 at the Nationa Cancer Institute, Milan

	Total	Adults	Children
Patients entered on study	483	394	89
Patients evaluable	410	327	83
Pretreated patients	149	128	21
Untreated patients	261	199	62
Adriamycin administered alone	296	252	44
Adriamycin+VCR+Pred. (leukemias)	21	10	11
Adriamycin+VCR+EDX+MTX (solid tumors)	24	—	24
Adriamycin+VCR+HN2+Pred. (lymphomas)	35	31	4
Adriamycin+VCR+HN2+BLM+Pred. (lymphomas)	34	34	—

VCR: vincristine EDX: cyclophosphamide MTX: Methotrexate
BLM: bleomycin HN2: nitrogen mustard Pred.: predinsone

Dosage

Table 2 reports the five dose schedules employed during Phase I and Phase II studies. Schedule E is still under evaluation in co-operation with the South West Cancer Chemotherapy Study Group (S. W. C. C. S. G.) of USA. Fig. 1-4 summarize the present protocols of intensive treatment for acute leukemia ,malignant lymphomas and some solid tumors in children where adriamycin is given in combination with conventional drugs and methods of therapy. A detailed description of these protocols is beyond the scope of this paper.

Table 2. Dose-schedules employed during Phase I and II evaluation of adriamycin (296 cases)

Dose-schedules	Description
A 0.4-0.65-0.8 mg/kg	Drug injected for 4 days, then stopped for 3 days. Restarted once or twice a week.
B 0.4-0.65-0.8 mg/kg	Drug injected every other day for 4-6 doses, then stopped for 3 days. Restarted once or twice a week.
C 0.4-0.65-0.8 mg/kg	Drug injected for 3 days. Course repeated after a 4 day rest period. After a second 4 day rest period the drug was restarted once or twice a week.
D 0.4-0.6 mg/kg	Similar to schedule C except for the rest period: 7 days.
E 20-25 mg/m^2	Drug injected for 3 days. Course repeated every 3 weeks. Dosage adjusted according to leukocyte and platelet levels[a].

[a] In cooperation with the S.W.C.C.S.G. (USA).

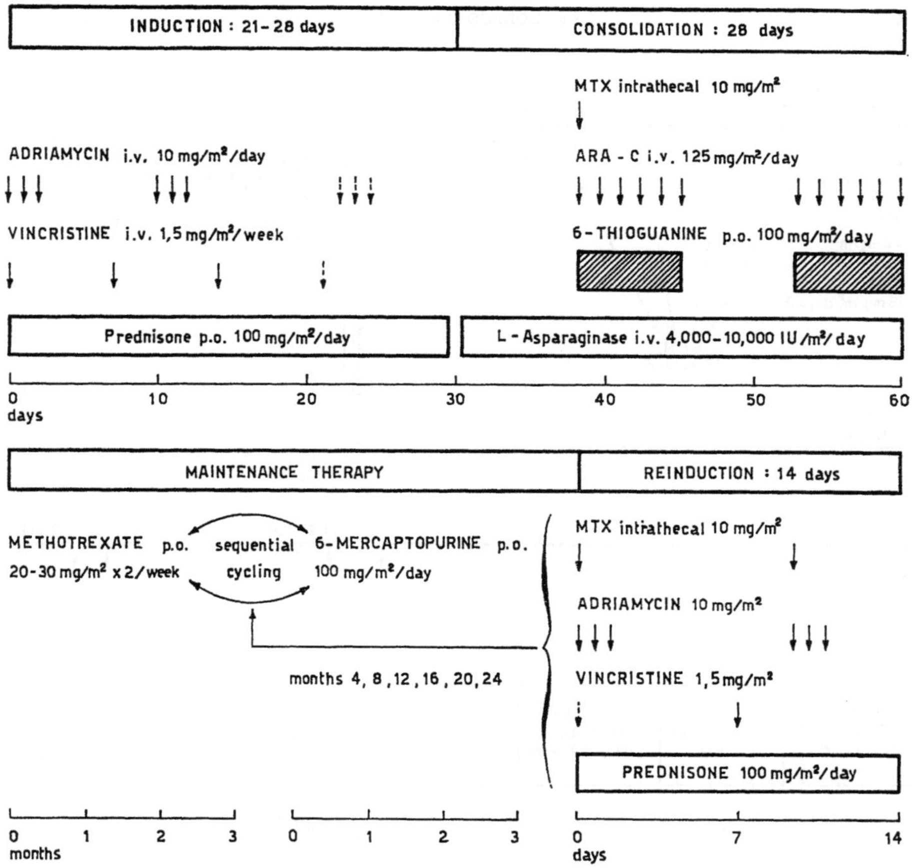

Fig. 1. Intensive treatment for acute lymphoblastic leukemia, acute lymphosarcoma cell leukemia and blastic phase of chronic myeloid leukemia

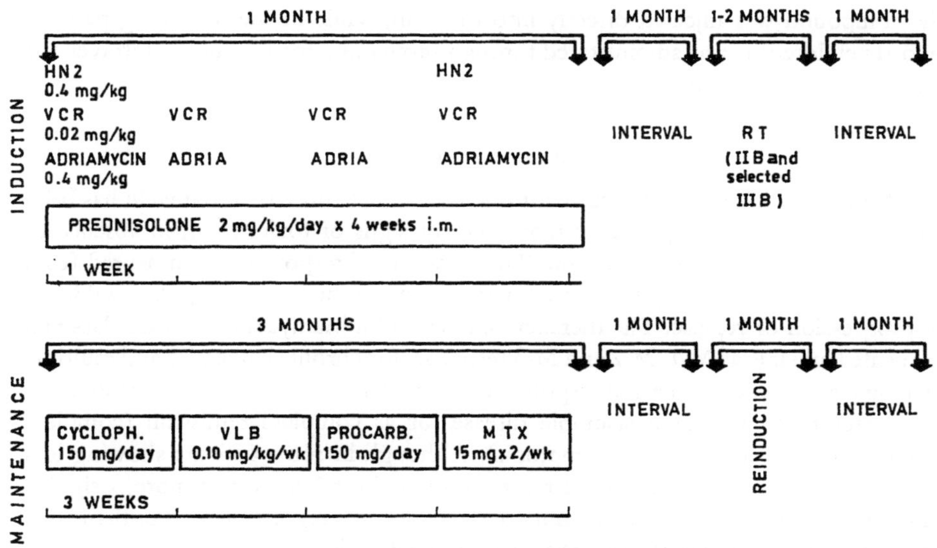

Fig. 2. Continuous combination chemotherapy in untreated malignant lymphomas (protocol 7009/a)

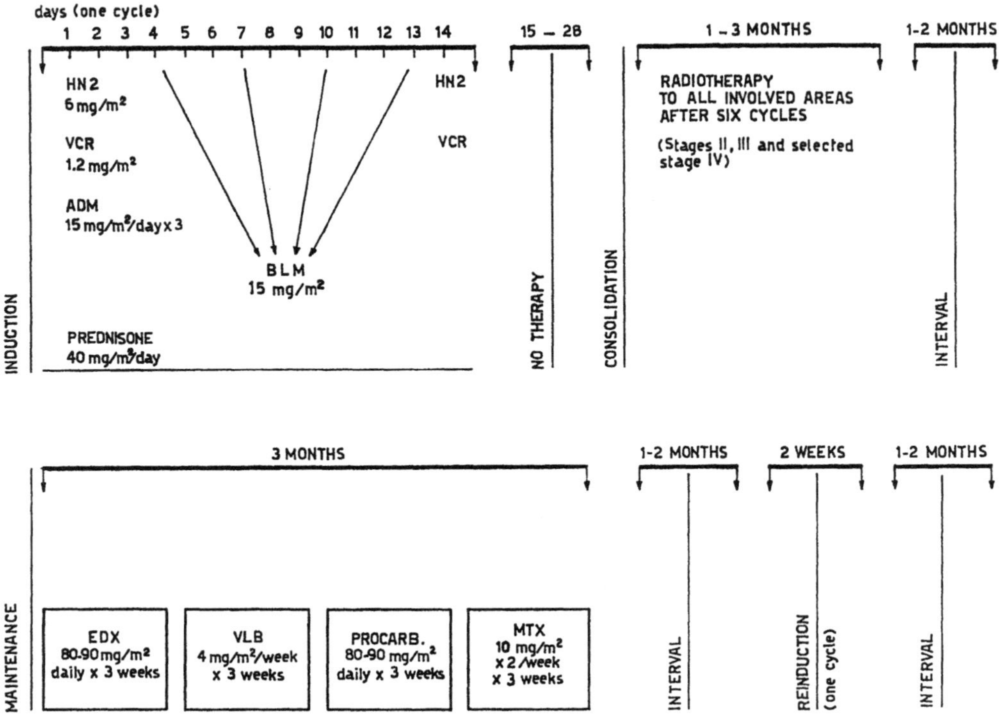

Fig. 3. Intensive treatment for malignant lymphomas (protocol 7009/b)

The drug, avialable as a reddish powder in 20 mg and more recently in 10 mg vials, has always been diluted in 0.9% NaCl solution. In about two thirds of cases the compound was injected directly into the vein, while in the remaining patients (mostly children) it was administered through tubing as a freely running intravenous infusion to avoid extravasation.

Evaluation of Response

During all Phase I and Phase II studies, an adequate dose of adriamycin has been defined as that producing one or more evident signs of toxicity, or, in absence of toxicity, a definite tumor regression. The duration of response was measured from the time of initial but unequivocal tumor regression to the time of objective evidence of progression or replase. The therapeutic effects of adriamycin have been classified for acute leukemia as M-1, M-2 and M-3 response, for chronic leukemias and multiple myeloma as complete or partial response, and for all solid tumors as regression lesser or greater than 50% in measurable disease, or as complete remission (complete disappearance of all signs and symptoms). Although in some patients showing less than 50% response the duration of regression lasted for 1 month or more, a significant response has been considered only that producing a regression greater than 50% or a complete remission of more than 1 month's duration.

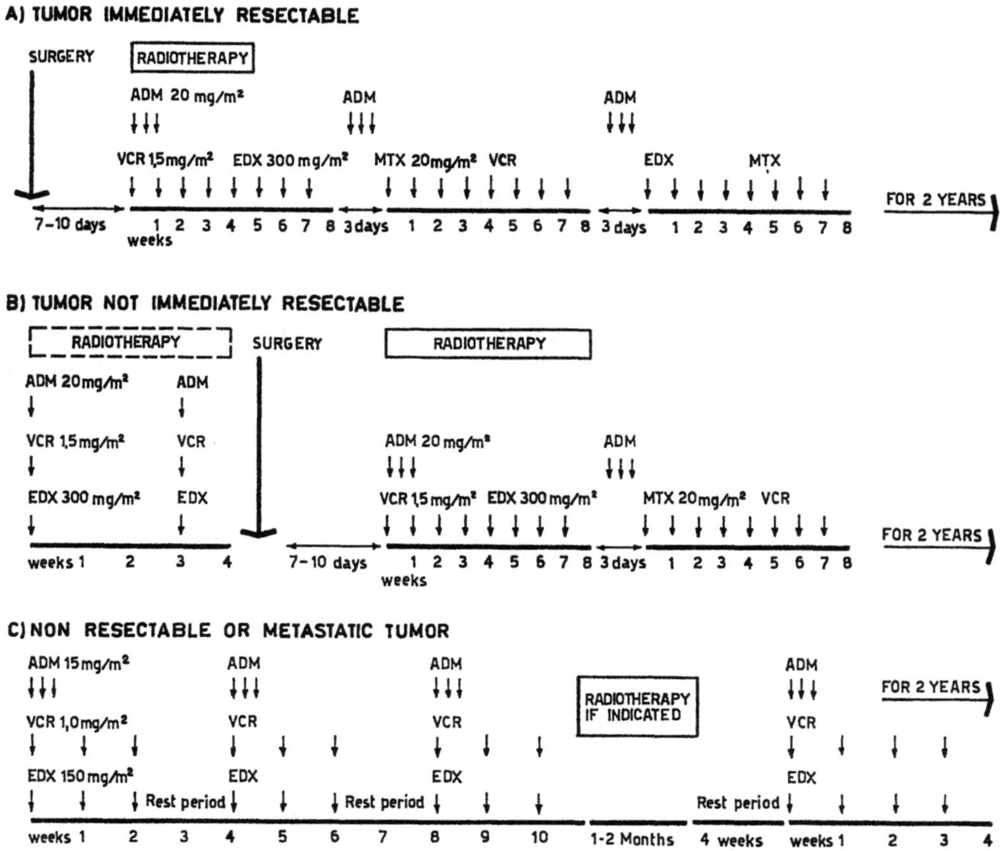

Fig. 4. Intensive treatment for solid tumors of children: soft tissue sarcomas, bone tumors, neuroblastoma

Results

Toxicity

The incidence of toxic manifestations in relation to each dose-schedule is summarized in Table 3. The detailed description of side effects has been reported in other

Table 3. Toxicity after different dose-schedules of adriamycin (296 evaluable cases)

Side Effects	Dose-schedules A (29 cases) (in %)	B (18 cases) (in %)	C (96 cases) (in %)	D (102 cases) (in %)	E (51 cases) (in %)
Alopecia	85	87	90	96	94
Stomatitis	74	75	79	47	37
Myelosuppression	81	68	84	81	45
Gastroenteritis	22	12	36	14	21
Cardiac toxicity	none	6.2	3.2	2	1.9
No toxicity	none	6	1	3	1.9

About 15% of patients showed fever and 5% chemical cellulitis due to local extravasation.

publications [10, 11, 24]. The most consistent side effects are represented by a triad: alopecia, stomatitis and myelosupression. Loss of hair to complete alopecia occurred in almost all patients, but particularly in those given the drug with an intermittent dosage (schedules D and E) which allowed in most cases the administration of a higher total dose of adriamycin. With schedule E the incidence of stomatitis and of myelosuppression dropped to values about 50% of those observed with previous regimens, except for oral mucositis in schedule D. Furthermore, the degree of both stomatitis and of bone marrow depression was usually less severe with schedule E than when giving the drug with the other schedules. In particular, no patients developed severe and prolonged agranulocytosis. Reversible electrocardiographic (ECG) changes were recorded in about 30% of patients and almost always in adults. The results of serial ECG's taken during the course of therapy in the first 131 cases are reported in Table 4. As can be seen, an almost equal number of patients before starting adriamycin had either a normal ECG or different types of abnormalities,

Table 4. Electrocardiographic findings observed in the first 131 patients where the ECG was systematically recorded before, during treatment with adriamycin and about one week after the last dose

ECG. before adriamycin	cases	ECG after adriamycin	cases
Within normal limits	68	Unchanged	44
		Abnormal	14
		Transitory abnormalities	10
With different types of abnormalities	63	Unchanged	20
		Improved	13
		Increased abnormalities	21
		Transitory worsening	9

In 7/403 (1.7%) patients the clinical cause of death could be related to cardiac toxicity.

characterized in most cases by tachycardia, ventricular extrasystoles, depression of S-T segment, and flattening or inversion of T waves, especially in the left leads. In the majority of patients the abnormal ECG pattern was probably due to widespread disease with diffuse pulmonary metastases, to moderate or severe anemia, and occasionally to some electrolyte disturbances. Very few patients had a past history of cardiovascular disease. Fatal cardiac toxicity was suspected in 7/403 patients (1.7%). As already reported with daunomycin [8, 13, 20], the presumptive clinical cause of death was acute forward or congestive cardiac failure occurring within 12–48 h. Severe tachycardia and ECG signs of myocardial ischemia were present in the left leads, associated with a rapid fall in the arterial blood pressure. Post mortem examination performed in 5/7 patients failed to detect findings suggestive of morphological injury due to cumulative drug toxicity. Fatal cardiac toxicity was abrupt, it occurred only in adults (average age 48.5 years) and was not apparently related to the total dose administered in patients older than 60 years. In younger patients on the contrary symptomatic cardiac toxicity, occurred with total doses higher than 600 mg/m² (schedule E). This complication was not observed in the group of patients given adriamycin in combination (none has received doses higher than 600 mg/m²).

No significant changes in serum electrolytes, blood urea nitrogen, fasting blood sugar, liver function tests, and serum electrophoresis which could be attributed to drug effects were observed. A rise in the level ov serum uric acid was seen in some cases with malignant lymphomas or leukemia. However, it should be recalled that most patients, and in particular those with acute or chronic leukemia, received allopurinol during the entire period of drug administration.

Response

The therapeutic results after administration of adriamycin alone and in combination in various neoplastic diseases of adults and children are summarized in Tables 5—14. The highest percentage of response to adriamycin given alone (significant regression rate ranging from 23—40%) was observed in malignant lymphomas (Table 5), in soft tissue sarcomas (Table 6) and in miscellaneous solid tumors of

Table 5. Therapeutic effects of adriamycin given alone in malignant lymphomas (57 adults and 6 children)[a]

Histologic type	Cases	Response				Regression rate	
		None	<50%	>50%	Complete	Significant (in %)	Overall (in %)
Hodgkin's disease	28	5	13	7	3	35%	82%
Reticulum cell sarcoma[b]	20 (2)	3 (1)	5	11 (1)	1	60%	85%
Lymphosarcoma[b]	15 (4)	2 (1)	10 (3)		3	20%	86%
Total	63 (6)	10 (2)	28 (3)	18 (1)	7	40%	84%

[a] In parentheses the number of children.
[b] Patients with bone marrow and blood invasion are excluded.

Table 6. Therapeutic effects of adriamycin given alone in soft tissue sarcomas of adults and children (33 adults and 6 children)[a]

Histologic type	Cases	Response				Regression rate	
		None	<50%	>50%	Complete	Significant (in %)	Overall (in %)
Fibrosarcoma	10 (2)	6 (1)	2	2	(1)		
Liposarcoma	4		2	2			
Rhabdomyosarcoma	7 (1)	2	4	1 (1)			
Angiosarcoma	4 (1)	1	2		1 (1)		
Myxosarcoma	1		1				
Synovialsarcoma	2 (1)	1	1 (1)				
Hermangiopericytoma	1		1				
Undifferentiated Sa.	10 (1)	3	4	3 (1)			
Total	39 (6)	13 (1)	17 (1)	8 (3)	1 (1)	23%	65%

[a] In parentheses the number of children.

Table 7. Therapeutic effects of adriamycin given alone in miscellaneous solid tumors of children (28 cases)

Histologic type	cases	Response				Regression rate	
		None	<50%	>50%	Complete	Significant (in %)	Overall (in %)
Neuroblastoma	11	2	4	4	1		
Wilms' tumor	5		3	2			
Osteogenic sarcoma	3	3					
Ewing's sarcoma	2		1	1			
Chondrosarcoma	1	1					
Teratosarcoma testicle	1		1				
Letterer-Siwe disease	1		1				
Neuroepithelioma	1		1				
Ameloblastoma	1		1				
Hepatoma	2	1	1				
Total	28	7	13	7	1	27%	75%

Table 8. Therapeutic effects of adriamycin given alone in testicular tumors (20 cases)

Histologic type	Cases	Response				Regression rate	
		None	<50%	>50%	Complete	Significant (in %)	Overall (in %)
Seminoma	4	2	1	1			
Embryonal Ca	10	4	5	1			
Terato Ca	5		5				
Chorio Ca	1			1			
Total	20	6	11	3		15%	70%

Table 9. Therapeutic effects of adriamycin given alone in lung carcinomas (25 cases)

Histologic type	Cases	Response			Rgresseion rate	
		None	<50%	>50%	Significant (in %)	Overall (in %)
Epidermoid Ca	5		5			
Adenocarcinoma	6	4	1	1		
Oat cell Ca	4	1	1	2		
Undifferentiated	5	3	2			
Unknown	5	1	3	1		
Total	25	9	12	4	16%	64%

children (Table 7). Tumor regression was seen in numerous histologic expressions of neoplastic disease, but particularly in all types of malignant lymphomas, in embryonal and teratocarcinomas of testicle (Table 8), in epidermoid and oat-cell carcinomas of lung (Table 9), in liposarcoma, rhabdomyosarcoma and angiosarcoma, in neuroblastoma, Wilms' tumor and Ewing's sarcoma. The response rate as well as the number of complete remissions was obvioulsy higher when the drug was administered in combination with other agents (Tables 10 and 11).

Encouraging results were also observed in a limited series of cases of acute lymphoblastic leukemia in children and in acute lymphosarcoma cell leukemia (M-1 response in 3/5 cases given adriamycin alone and in 9/12 cases given adriamycin in combination with vincristine and prednisone) (Table 12). Adriamycin was moderately effective in 5 patients with chronic myelogenous leukemia, in 7 patients with chronic lymphocytic leukemia and in 4 with multiple myeloma (Table 13). A number of regressions was also observed in a group of miscellaneous solid tumors of adults (Table 14). Results were promising in carcinoma of the breast, in carcinomas of the gastrointestinal tract, in carcinoma of the nasopharynx, in Ewing's sarcoma, in chorioepithelioma of the uterus and in transitional carcinoma of urinary bladder.

Regressions were usually prompt occurred within 2—3 weeks in the large majority of patients. However, they were short-lived in most cases (1.5—4.5 months) with or without maintenance therapy. Nevertheless, it should be noted that the administration of adriamycin alone was able to control the neoplastic disease for at least 6 months in 10 patients and for 1 year or more in 8 patients (Table 15). The prolonged regression was observed in a fairly large variety of tumors and 6/18 cases showed a complete remission.

Table 10. Therapeutic effects of intensive combination chemotherapy including adriamycin in malignant lymphomas (65 adults and 4 children)[a]

Histologic type	Cases	Quadruple therapy				Significant	Cases	Quintuple therapy[c]				Significant
		None	< 50%	> 50%	Complete	Regression rate (in %)		None	< 50%	> 50%	Complete	Regression rate (in %)
Hodgkin's disease	17 (3)		3	2	12 (3)	82%	13		2	8	3	84%
Reticulum cell Sarcoma[b]	8		3	2	3	62%	11		2	7	2	81%
Lymphosarcoma[b]	10 (1)	1 (1)	3	1	5	60%	10		3	6	7	70%
Total	35 (1)	1 (1)	9	5	20 (3)	71%	34		7	21	6	79%

[a] In parentheses the number of children.
[b] Patients with bone marrow and blood invasion are excluded.
[c] 19 patients are still on therapy (induction phase).

Table 11. Therapeutic effects of adriamycin given in combination with vincristine, cyclophosphamide and methotrexate in solid tumors of children (24 cases)

Histologic type	Cases	Response				Precautional therapy
		None	<50%	>50%	Complete	
Neuroblastoma	10	1	4 (1)	4 (1)		1
Rhabdomyosarcoma	5			1	1 (1)	3
Soft tissue sarcomas	3					3
Ewing's sarcoma	4		2		2 (2)	
Chondrosarcoma	1					1
Osteosarcoma	1		1 (1)			
Total	24 (6)	1	7 (2)	5 (1)	3 (3)	8

In parentheses the number of patients receiving radiotherapy with chemotherapy.

Table 12. Therapeutic effects of adriamycin in acute leukemia (16 children and 19 adults)

Cytologic type	Cases	Response to adriamycin alone			Cases	Response to adriamycin in combination		
		M–1	M–2	M–3		M–1	M–2	M–3
				Cildren				
ALL	3	2	1		6	5	1	
AC. LSA cell leukemia	2	1		1	5	5		
Total	5	3	1	1	11	10	1	
				Adults				
ALL	4		1	3				
AML	3		1	2	1			1
AEL[a]	1		1					
AC. LSA cell leukemia	2		1	1	6	2	3	1
CML (blastic)	1			1	1	1		
Total	11		4	7	8	3	3	2

[a] Acute erythroleukemia.

Discussion

During the past 3 years the new antitumor antibiotic adriamycin has received considerable attention in numerous centers throughout the world. The available reports from different investigators [14, 21—23, 27—30, 32, 33] confirm that this drug shows definite therapeutic activity against several types of neoplastic disease in children and adults. In particular, there seems to be evidence that adriamycin is equally as effective as daunomycin in acute lymphoblastic leukemia of children [28—32]. In

Table 13. Therapeutic effects of adriamycin in chronic leukemias (17 cases) and in multiple myeloma (7 cases)

Cytologic type	Cases	Response to Adriamycin alone			Cases	Response to adriamycin in combination[a]		
		Complete	Partial	None		Complete	Partial	None
CML	7	2	4	1				
CLL	10	1	6	3	2		2	
Total	17	3	10	4	2		2	
Multiple Myeloma	7		4	3				

[a] Adriamycin + Vincristine + Prednisone.

Table 14. Therapeutic effects of adriamycin given alone in miscellaneous solid tumors of adults (88 cases)

Histologic type	Cases	Response			Regression rate	
		None	$<50\%$	$>50\%$	Significant (in %)	Overall (in %)
Ca breast	19	9	7	3		
Ca G.I. traat	10	2	6	2		
Ca nasopharynx	8	1	5	2		
Ca oral cavity	8	5	2	1		
Ca primary site unknown	7	4	2	1		
Malignant melanoma	6	6				
Ewing's sarcoma	5	1	1	3		
Ca thyroid	3	2		1		
Ca maxillary sinus	2	1	1			
Chorioepithelioma uterus	2		1	1		
Osteogenic sarcoma	2	1	1			
Ca urinary bladder	4	1	2	1		
Ca Kidney	2	1	1			
Ca parotid	1	1				
Ca cervix	1		1			
Dysgerminoma ovary	1	1				
Ca skin	1		1			
Chondrosarcoma	1	1				
Malignant thymona	1	1				
Neuroepithelioma	1			1		
Paraganglioma	1		1			
Pheochromocytoma	1	1				
Sarcoma uterus	1	1				
Total	88	40	32	16	18%	54%

Table 15. Prolonged significant remissions after treatment with adriamycin alone in adults and children (18/289 cases)

Hystologic type	Cases	Duration of Remission	
		> 6 months	1 year or more
Hodgkin's Disease	4	3	1 (CR)
Lymphosarcoma	1		1
Reticulum cell Sa.	1		1 (CR)
Neuroblastoma	2	2 (CR 1)	
Liposarcoma	1	1	
Angiosarcoma	1		1 (CR)
Ewing's sarcoma	2	2	
Seminoma	1		1
Terato Ca. testicle	1		1
Ca. breast	1		1 (CR)
Ca. intestine	1	1	
Ca. lung (oat-cell)	1	1	
Ca. nasopharynx	1		1 (CR)
Total	18	10 (CR 1)	8 (CR 5)

CR: complete remission.

solid, tumors, the compound has been producing consistent regressions which, in most cases, are undoubted superior to those reported so far with daunomycin. Whe—
most cases, are undoubted superior to those reported so far with daunomycin.
Whether the therapeutic effectiveness of adriamycin is better than that of other active agents against a specific type of untreated neoplastic disease remains to be confirmed through appropriate clinical trials. Adriamycin does not seem to show cross-resistance with radiation therapy or with most conventional drugs, with the exception of daunomycin, although this specific problem requires a more detailed evaluation, especially in patients with acute leukemia. Toxicity can be in part explained by the results of studies on the physiologic disposition of ^{3}H-adriamycin. The pattern of distribution and excretion appears from preliminary studies [16] to be similar to that of daunomycin [15]. The compound is rapidly fixed by body tissues while moderate to low urinary and fecal excretion persists for about 2 weeks from intravenous or intracavitary injection. Prolonged daily administration can, therefore, produce accumulation and early toxicity. As far as our clinical experience is concerned, several conclusions can be drawn. They can be summarized as follows:

1. Adriamycin is a potent new antitumor antibiotic with a higher therapeutic index than daunomycin and effective on a large variety of neoplastic diseases.

2. The incidence of toxic signs, namely stomatitis and myelosuppression, can be definitely decreased by an intermittent dose-schedule (interval of 3 weeks between courses), without affecting the therapeutic potential.

3. Adriamycin produced fatal cardiac toxicity in 1.7% of cases. This did not occur in patients given adriamycin in combination. Therefore, it is possible that small intermittent doses, as given in association with other drugs, can lower if not avoid this complication. Whenever possible, a total dose higher than 600 mg/m^2 should be

avoided. This is probably the dose-limit for symptomatic cardiac toxicity. Serial ECG can monitor myocardial damage, but usually are not a reliable diagnostic tool in patients with widespread cancer. In any case, it is our opinion that the risk of this undesiderable manifestation appears acceptable when compared with the significant therapeutic benefits of adriamycin.

4. Adriamycin administered alone at half the dose used with daunomycin was therapeutically active in acute lymphoblastic leukemia of children, lymphomas and particularly in a large variety of solid tumors of children and adults. Regressions were prompt but usually short-lived, indicating that the administration of adriamycin alone seems more useful in inducing the initial tumor regression than as maintenance therapy.

5. In those types of neoplastic disease where the drug proved to be significantly effective, adriamycin should be tried in combination with conventional agents.

In conclusion, our data indicate that adriamycin can be considered an additional modality of antitumor therapy for use in patients with lymphoblastic leukemia and solid tumors, and thus may expand the therapeutic measures for neoplastic disease.

Acknowledgements

The authors wish to thank the following physicians for their effective co-operation in the evaluation of patients during the trials with adriamycin:

G. M. De Palo, A. Cunsolo, C. Bartoli, E. Bajetta, G. Tancini, A. Guindani, F. Pizzetti, and S. Di Pietro.

The authors are indebted to Mr. S. Ceccarini for his skillful technical assistance in the preparation of the graphs.

References

1. Arcamone, F., Franceschi, G., Tenco, S., Selva, A.: Tetrahedron Letters **13**, 1007-1010 (1969).
2. Bonadonna, G.: Valutazione clinica dell' adriamicina. Corso Superiore Chemioter. Tumori, Milano, 23-27 March 1970, CEA, Milano, pp. 357-367.
3. Bonadonna, G.: Clinical trials with adriamycin in the neoplastic disease of children and adults. Proc. 7° Int. Congress Chemother., Prague, 23-28 August 1971 (Panel discussion).
4. Bonadonna, G., Beretta, G.: Proc. Soc. Ital. Cancer. **7**, 279-290, (1970) CEA, Milano.
5. Bonadonna, G., De Lena, M., Beretta, G.: Europ. J. Cancer. **7**, 365-367 (1971).
6. Bonadonna, G., De Lena, M., Beretta, G., Fossati-Bellani, F., Monfardini, S.: Treatment of leukemias and lymphomas with adriamycin. Proc. 13° Int. Congress Hemat., Munich, 2-8 August 1970, p. 96.
7. Bonadonna, G., Fossati-Bellani, F., De Lena, M., Monfardini, S., Beretta, G., Bajetta, E.: Proc. Amer. Soc. Clin. Oncology **7**, 3, 1971 (Abstr. 9).
8. Bonadonna, G., Monfardini, S.: Cardiac Toxicity of Daunorubicin. Lancet 1969 I, 837.
9. Bonadonna, G., Monfardini, S.: Proc. Amer. Ass. Cancer Res. **11**, 10 (1970) (Abstr. 37).
10. Bonadonna, G., Monfardini, S., De Lena, M., Fossati-Bellani, F.: Brit. Med. J. 1969 III, 503-506.
11. Bonadonna, G., Monfardini, S., De Lena, M., Fossati-Belani, F., Beretta, G.: Cancer Res. **30**, 2572-2582 (1970).
12. Bonadonna, G., Monfardini, S., Di Pietro, S.: Clinical evaluation of adriamycin. Proc. 10° Int. Cancer Congress, Houston, 22-29 May, 1970 (Abstr. 774)
13. Bonadonna, G., Monfardini, S., Guindani, A.: Tumori **6**, 465-480 (1968).
14. De Cataldo, F., Praga, C., Bussi, L., Polli, E.: Adriamycin in acute leukemia. Proc. 13° Int. Congress Hemat., Munich 2-8 August, 1970, p. 95.
15. Di Fronzo, G., Bonadonna, G.: Europ. J. Clin. Biol. Res. **15**, 314-320 (1970).

16. Di Fronzo, G., Gambetta, R., Bonadonna, G.: Preliminary studies on the distribution and excretion of ^{3}H-adriamycin in man. Proc. 1° Congress Europ. Ass. Cancer Res., Bruxelles 14-17 September 1971, p. 25.
17. Di Marco, A., Gaetani, M., Scarpinato, B.: Cancer Chemother. Rep. **53**, 33-37 (1969).
18. Di Pietro, S., De Palo, G. M., Molinari, E., Gennari, L.: Tumori **56**, 233-244 (1970).
19. Karnofsky, D. A., Burchenal, J. H.: The clinical evaluation of chemotherapeutic agents in cancer. In: MacLeod, C. M. (Ed.): Evaluation of chemotherapeutic agents. New York: 1949 Columbia Univ. Press.
20. Marmont, A. M., Damasio, E., Rossi, F.: Lancet 1969 I, 837.
21. Massimo, L., Cottafava, F., Mori, P. G., Fossati-Guglielmoni, A.: Minerva Pediat. **21**, 2182-2186 (1969).
22. Middleman, E., Luce, J., Frei, E. III: Proc. Amer. Soc. Clin. Oncology **7**, 10, (1971) (Abstr. 39)
23. Mathé, G., Amiel, J. L., Hayat, M., De Vassal, F., Schwarzenberg, L., Schneider, M., Jasmin, C., Rosenfeld, C.: Presse med. **78**, 1977-1999 (1970).
24. Monfardini, S., Bonadonna, G., Di Pietro, S., Guindani, A., Fossati-Bellani, F., De Lena, M.: Tumori **55**, 197-216 (1969).
25. Monfardini, S., De Lena, M., Fossati-Bellani, F., Beretta, G., Bonadonna, G.: Therapeutic effects of adriamycin in leukemias and in lymphomas. Proc. 2° Meeting Asian Pacific Div. Int. Soc. Hemat., Melbourne 24-28 May, 1971.
26. Monfardini, S., Bajetta, E., Musumeci, R., Bonadonna, G.: J. Urol., 1972 (in press).
27. Polli, E., Praga, C.: Adriamycin in the treatment of inoperable lung tumor. Proc. 7° Int. Congress Chemother., Prague, 23-28 August (1971) (Abstr. B-7/16).
28. Tan, C., Wollner, N., Haghbin, M.: Clinical trials of adriamycin. Proc. 10° Int. Cancer Congress, Houston 22-29 May (1970) (Abstr. 775).
29. Tan, C., Wollner, N., King, O., Ilang, D.: Proc. Amer. Ass. Cancer Res. **11**, 79 (1970) (Abstr. 311).
30. Tan, C., Wollner, N., Murphy, M. L.: Adriamycin in children with acute leukemia and lymphoma. Proc. 13° Int. Congress Hemat., Munich, 2-8 August, 1970, p. 96.
31. Tan, C., Tasaka, H., Yu, K.-P., Murphy, L., Karnofsky, D. A.: Cancer **20**, 333-353 (1967).
32. Wollner, N., Tan, C., Ghavimi, F., Rosen, G., Teff, M., Murphy, M. L.: Proc. Amer. Ass. Cancer Res. **12**, 75 (1971) (Abstr. 300).
33. Wang, J., Cortes, E., Skins, L., Holland, J. F.: Proc. Amer. Soc. Clin. Oncology **7**, 14 (1971) (Abstr. 54).

Clinical Trials of Adriamycin

Emil Frei, III, James K. Luce, and Edward Middleman

Department of Developmental Therapeutics, the University of Texas
M. D. Anderson Hospital & Tumor Institute at Houston
Houston, Texas 77025

With 3 Figures

Introduction

Adriamycin is a tetracyclic anthracycline antibiotic with substantial antitumor activity in transplanted rodent tumors. It is a 14-hydroxy derivative and thus a close analog of daunorubicin. Its spectrum of antitumor activity in experimental systems is similar to that of daunorubicin though the therapeutic index of adriamycin is generally superior to that of daunorubicin. The tolerated dose of adriamycin in experimental systems is generally somewhat less (approximately 2/3) that of daunorubicin. In pharmacologic studies the anthracyclines bind to macromolecules particularly DNA and are slowly excreted. Whereas daunorubicin is extensively metabolized adriamycin is minimally metabolized. This probably accounts for the greater toxicity of a given dose of adriamycin.

Dose Schedules

The dose schedules employed in our studies are delineated in Table 1. The first and third dose schedule involved a 3 day course and a 1 day course respectively given at 3 week intervals. The second program involved weekly single dose treatment. It should be noted that the total dose tolerated per unit time was comparable for the 3 treatment programs. This is similar to experience with alkylating agents and with compounds that are tightly bound to marcomolecules. It contrasts with many of the

Table 1. Adriamycin dose schedules

Dose schedule producing equivalent toxicity	Total dose (mg/m^2)
1) 20- 30 $mg/m^2/d \times 3$; q 3 wks. $\times$ 3	180-270
2) 20- 35 mg/m^2 q wk. $\times$ 9	180-315
3) 60-105/m^2 q 3 wk. $\times$ 3	180-315

antimetabolites such as arabinosyl cytosine and methotrexate where continuous treatment results in a much smaller tolerated dose than intermittent treatment.

In experimental systems there is little evidence for schedule "sensitivity" for adriamycin. In GOLDIN's study in L1210 the most effective schedule was treatment given every 3 h for 24 h. While such treatment provided a somewhat longer survival, host recovery was delayed and second courses of treatment were difficult to deliver.

In our clinical studies we are increasingly emphasizing intermittent intensive treatment as compared to continuous treatment. Thus in experimental systems intermittent treatment tends to be superior and recent data in Burkitt's lymphoma, acute leukemia, multiple myeloma, lymphoma and choriocarcinoma strongly suggest the superiority of intermittent intensive treatment to continuous daily treatment. In addition to pharmacologic and cytokinetic bases for such intermittent treatment a major factor relates to immunosuppression.

In Fig. 1, the effects of intensive 5 day courses of treatment with antitumor agents on immunosuppression is presented for 3 parameters of the immune response:

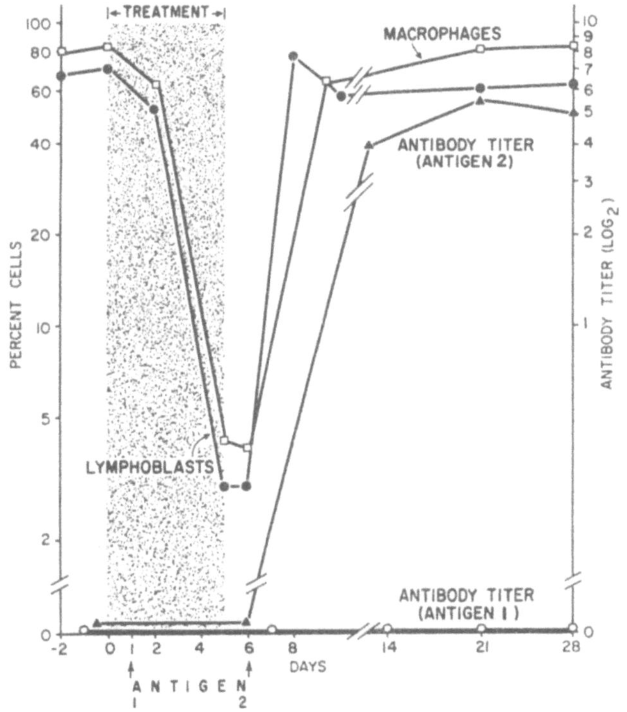

Fig. 1

macrophage ingress into skin windows, circulating antibody response to primary antigenic stimulation and lymphocyte blastogenesis with PHA. There is a marked reduction in immune response during treatment but rapid recovery in the 3—5 days following treatment. Thus if treatment is given every 3 weeks the patient is immunosuppressed for a small part of the time. In contrast, in studies of continuous daily treatment immunosuppression occurs which initially is not as severe as that described above but tends to be progressive. Thus immunologically there are 3 partial advantages to intermittent treatment 1) opportunistic infections in patients on long term treatment are less likely to occur, 2) immune restraint of the tumor is less impaired and 3) an active immunotherapy program can be properly timed into chemotherapy regimens.

In Fig.s. 2 and 3, the effect of 3 day courses of adriamycin at 30 mg/m²/day at 3 week intervals on the peripheral leukocyte and platelet count respectively are presented. Granulocytopenia of moderate severity occurs regularly and thrombocytopenia is less severe. Recovery is generally complete by 3 weeks and such courses

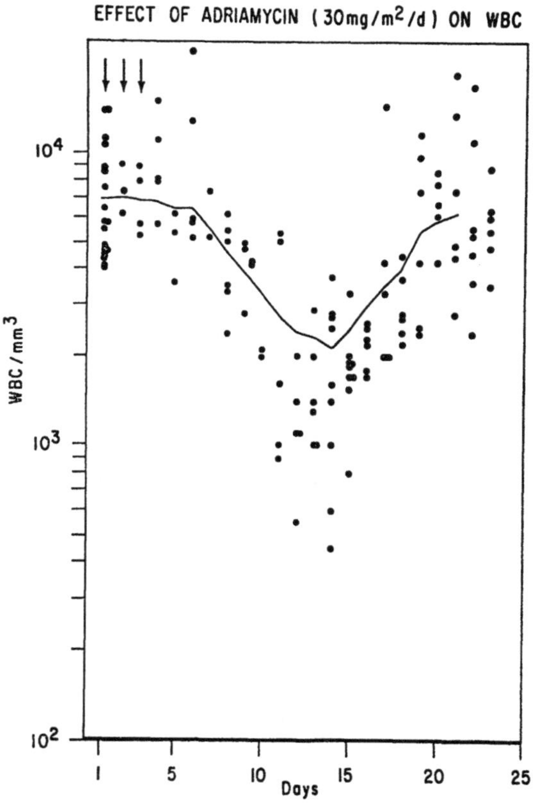

Fig. 2. Effect of adriamycin (30 mg/m²/d) on WBC

can be repeated at 3 week intervals without cumulative toxicity. Thus the evolution of myelosupression is comparable to that of cyclophosphamide. Marrow recovery tends to occur more rapidly with cell cycle specific agents such as ara-C and hydroxyurea. On the other hand marrow depression is more prolonged with the nitrosoureas.

The effect of the dose of adriamycin on leukopenia and thrombocytopenia is presented in Table 2. A dose response relationship is apparent this is also true for mucositis as presented in Table 3.

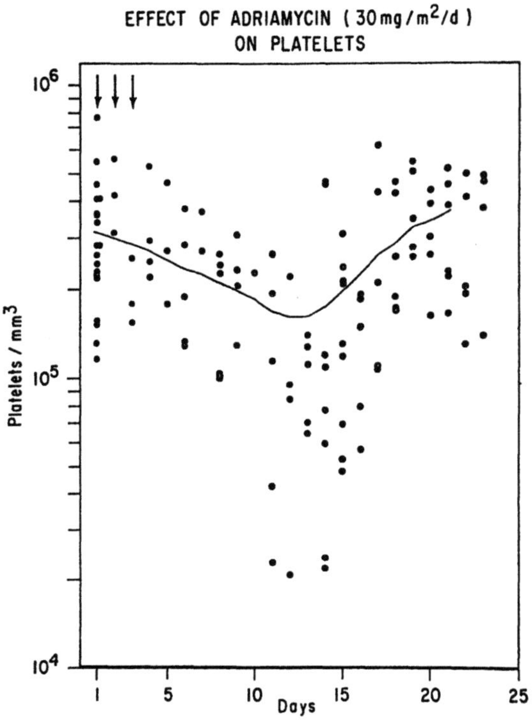

Fig. 3. Effect of adriamycin (30 mg/m²/d) on platelets

Table 2. White blood cell and platelet response to adriamycin

Dose and schedule (mg/m²/day)	Total dose (mg/m²)	Number evaluable patients (WBC/platelets)	WBC		Platelets	
			Count (mm³)	Day of nadir	Count (x 10³/mm³)	Day of nadir
18 × 5d	90	7/ 8	2660	17	200	13
30 × 3d	90	20/19	2150	14	164	12
25 × 3d	75	5/ 4	3035	14	217	12
20 × 3d	60	9/ 6	3475	16	317	15

Table 3. Incidence of mucositis in patients receiving adriamycin

Total dose per course	First course	Second through fifth course
60- 75 mg/m²	20% (3 /15)	9.5% (4/42)
90-112 mg/m²	52% (16/31)	88 % (35/40)

Toxicity

As above, myelosuppression was dose related and was the major dose limiting toxic effect. Some degree of alopecia (often severe) occured in 90% of patients and cessation of beard growth regularly occurred. Nausea, vomiting and diarrhea occurred in approximately 20% of patients. All of the above toxic effects were related to total dose but were independent of schedule. An exception was oral ulceration. A few patients were treated with 5 day courses of adriamycin given at 8 h intervals. Mucositis, often severe, occurred in all of these patients. Mucositis occurred in 60% of patients receiving 3 day courses of treatment but in only 10% of patients receiving a single dose every 3 weeks. The reason for this difference in mucositis and the lack of difference in the other toxic manifestations is not known.

The tolerated dose ratio of daunorubicin to adriamycin is approximately 3:2. This corresponds closely to the toxicity ratio for the 2 agents in experimental systems, particularly the rodent.

Antitumor Effect

There were too few patients in any given tumor category treated with the 3 dose schedules listed in Table 1 to draw conclusions with respect to the effect of dose schedule on antitumor response. The antitumor effects of adriamycin are presented in the next 6 figures. In Table 4 the effect of adriamycin on miscellaneous tumors is presented. "Evaluable" is defined as patients who had measurable tumor masses and who were treated with at least two courses of adriamycin. Complete remission (CR) constituted complete disappearance of objective evidence of disease. Partial remission (PR) was defined as a 50% or greater reduction in tumor mass measured, where possible, as a decrease in the product of 2 diameters. Of 11 evaluable patients with transitional cell carcinoma of the bladder there were 6 responses. Two of these patients had pulmonary metastases and the remainder had pelvic and/or liver lesions which could be measured with reasonable accuracy.

Table 4. Adriamycin and Miscellaneous tumors

Tumor	No. entered	No. evaluable	CR	PR	Total (%)
Total	58	49	4	9	13 (26)
Transitional Ca Bladder	15	11	3	3	6 (55)
Kidney	7	6	0	0	0
Testis	4	4	0	0	0
Melanoma	6	4			
Adrenal cortex	2	1	0	0	0
Thyroid Ad Ca	14	14	0	3	3 (22)
Astrocytoma	1	1	0	1	1
Lymphoepithelioma	1	1			
AML	5	4			
Hodgkin's	2	2	1	1	2 (100)
Lymphosa	1	1	0	1	1

Three of 14 patients with adenocarcinoma of the thyroid had partial responses. While only 3 patients with lymphoma have been treated all have responded including 1 complete remission. Response in lymphoma was rapid and two of the patients remained in remission 5 months after the start of adriamycin therapy.

In Table 5, the effect of adriamycin on adenocarcinoma of various sites is presented. Responses occurred only in patients with breast cancer where 3 of 4 patients had objective response. The patient who had a complete remission had complete disappearance of measureable pulmonary metastases which persisted for a period of 6 months.

A limited number of patients with squamous cell carcinoma were treated (Table 6). One of 3 patients with carcinoma of the skin responded. This was a patient with extensive local recurrence from squamous cell carcinoma of the anus. A marked (90%) reduction in tumor mass occurred lasting for 7 months.

Table 5. Adriamycin and adenocarcinoma

Tumor	No. entered	No. evaluable	CR	PR	Total response (%)
Total	29	27	1	2	3 (11)
Breast	4	4	1	0	3 (75)
Lieng	1	1	0	0	0
Stomach	2	2	0	0	0
Liver	2	1			
Bliary tract	1	1			
Pancreas	2	2	0	0	0
Colon	8	8	0	0	0
Prostate	2	1	0	0	0
Ovary	5	5	0	0	0
Unknown	2	2	0	0	0

Table 6. Adriamycin and squamous cell carcinoma

	No. entered	No. evaluable	CR	PR	Total response (%)
Head and Neck	3	3	0	0	0
Lung	8	5	0	0	0
Skin	3	3	0	1	1
Unknown	1	1	0	0	0
Total	15	12	0	1	1 (8)

The effect of adriamycin on soft tissue sarcomas is presented in Table 7. Nine of 27 (33%) of patients had objective tumor regression which in 2 instances was complete. Though the number of patients in each subcategory is relatively small responses tend to be distributed through the variety of soft tissue sarcomas. This is perhaps consistent with recent viral and immunologic studies suggesting common causal and immunologic relationships. The presence of C-type particles in some of these lesions and the known production of soft tissue sarcoma in experimental animals by RNA oncogenic viruses is of interest in view of the presence of the reverse transcriptase in RNA oncogenic virus infected cells and the recent reports that the anthracycline group of drugs may inhibit RNA dependent DNA polymerase.

Table 7. Adriamycin and soft tissue sarcoma

	No. entered	No. evaluable	CR[a]	PR[b]	Total response (%)
Soft tissue sarcoma (total)	28	27	2	7	9 (33)
Osteogenic Sa	5	5		0	0 (0)
Chondro Sa	4	4		1	1 (25)
Lipo Sa	3	3	1	1	2 (67)
Synovial Sa	1	1		0	0 (0)
Leiomyo Sa	3	3		0	0 (0)
Fibro Sa	3	3		1	1 (33)
Undifferentiated Sa	9	8	1	4	4 (63)

[a] Complete remission.
[b] Partial remission.

Combination Chemotherapy

Combination studies with adriamycin and dimethyl imidazole carboxamide triazeno (DIC) were initiated for the following reasons: 1) both compounds showed activity in soft tissue sarcoma and 2) Dr. Goldin has demonstrated synergism for the 2 agents in L1210 mouse leukemia. Moreover he has found that the best schedule involves a single dose of adriamycin followed by daily doses of DIC. In this setting the animals tolerate doses of each agent approaching those tolerated by the agents alone. The clinical schedule involved a single dose of adriamycin followed by 4 daily doses of DIC and such 5 day courses were repeated at 3 week intervals (Table 8). While patients were initially treated at 45 mg/m^2 of adriamycin and 200 mg/m^2/day for DIC it was soon recognized that this was relatively non-toxic and the doses were increased respectively to 90 mg/m^2 and 350 mg/m^2/day. Thus, as in the mouse, so

Table 8. Combined adriamycin and DIC in soft tissue sarcomas

	No. patients	No. responses (%)	
Adriamycin	27	9	(33)
DIC	55	8	(15)
Adria + DIC	15	5	(33)

Treatment program
Course Adriamycin 45-90 mg/m^2 d 1
DIC 200-350 mg/m^2 days 2-5
Courses given at 3 wk. intervals.

also in man these agents can be given together at doses approximating those employed when the agents are used alone. This combined study is preliminary and, of 15 patients treated to date, 5 have responded. It is to be emphasized that only the last 3—4 have been treated at full doses. Whether the combination proves superior to either agent alone remains to be determined.

In Table 9, our total experience with adriamycin is presented. Of 104 evaluable patients 22% had objective response, The median duration of response was relative short ($3^1/_2$ months) but it is to be emphasized that many of the patients were not maintained on treatment because of the possible risk of myocardial toxicity.

Table 9. Adriamycin - total experience

No. entered	No. evaluable	CR	PR	Total (%)
121	104	7	16	23 (22)
Median Duration Response 3.5 mo. (unmaintained)				

Promising areas: bladder, lymphoma, soft tissue sarcoma, breast

Conclusion

We conclude that adriamycin is a promising agent for certain non-leukemic neoplastic diseases. While direct comparisons have not been made, adriamycin would appear to be superior to daunorubicin in the treatment of solid tumors in man as it is in the rodent. Areas of particular promise in our experience include bladder cancer, lymphoma, soft tissue sarcomas and breast cancer. We are increasing emphasis on adriamycin studies in these areas, either alone or in combination with other agents.

Preliminary Results of a Clinical Trial with Intermittent Doses of Adriamycin in Lung Cancer

Y. Kenis and J. Michel[1]

Cooperative Group for Clinical Trials in Lung Cancer[2]
(EORTC)

With 3 Figures

This preliminary report is based on the study of 27 patients with far-advanced lung cancer treated at the Jules Bordet Institute, as part of a cooperative trial of the Cooperative Group for Clinical Trials in Lung Cancer, of the European Organization for Research on the Treatment of Cancer (EORTC). The complete data of the whole group of patients will be published later.

The purpose of this trial (a type I trial according to the terminology of the EORTC) was to evaluate the effect of adriamycin administered intermittenly in patients with lung cancer[3]. The histological diagnosis was epidermoid carcinoma in 18 cases, and anaplastic (oat-cell) carcinoma in 9 cases. All patients had far-advanced cancer and were not suitable for surgery or radiation therapy, according to the usual criteria of the group [1, 2].

The objective criteria for evaluation of response were as follows:

1. *A partial regression* was recorded when the decrease in measurable lesions was greater than 50% in the product of two diameters without progression of any other lesion and without appearance of any new lesions, over a period of at least one month; or a marked and unequivocal decrease (3/4 or more) of non-measurable lesions.

2. *A non-total failure* was characterized by a decrease of the lesions as in 1, but over a period of less than 1 month; or a definite decrease in some lesions while others were progressing.

3. *A total failure* was defined when there was progression of disease, absence of response, or any response smaller than 1 or 2.

Subjective improvement alone was not considered a positive response. Patients who received less than 2 doses were considered as non-evaluable.

---

1 Service de Médecine et d'Investigation Clinique de l'Institut Jules Bordet, Centre des Tumeurs de l'Université Libre de Bruxelles (Belgium).

2 Chairman of the group: A. Meyer (Paris); Secretary: P. Lévy (Paris).

3 Adriamycin was generously supplied by Farmitalia (Milan).

Adriamycin was given intravenously at a fixed dose of 75 mg/m^2 every 3 weeks. This dose was chosen according to the pharmacological data obtained by Di Fronzo et al. [3] with tritiated adriamycin.

Before the treatment and before each dose, kidney and liver functions were tested by several biochemical determinations. ECG and a dosage of creatine phosphokinase (CPK) were performed at the same time. Hemoglobin, white cell and platelet counts were carried out at least once a week during the course of treatment. Delayed skin sensitivity was tested before the treatment and periodically thereafter with 10 I.U. of PPD tuberculin intradermally.

Results

Four patients with epidermoid carcinoma received only one injection and were considered as non-evaluable for the anti-tumor effect.

Seven partial remissions were observed: 5 in epidermoid carcinomas and 2 in anaplastic carcinomas. There were also 2 non-total failures among the patients with anaplastic carcinoma (one partial response during less than 1 month; one case with decrease of some subcutaneous metastases while other were unchanged or slightly progressing). Fourteen patients were classified as total failures.

The longest partial remission lasted 6 months. The duration of the other remissions was: 3 months in 2 cases, 2 months in 3 cases, and 1 month in 1 case. Different localizations of the malignant disease responded to the treatment: the primary tumor (3 cases); subcutaneous nodules (2 cases); peripheral lymphadenopathies (2 cases); a pleural effusion and a superior vena cava syndrome (1 case). Three responders died from cerebral metastases without evidence of progression of the other lesions.

The median survival time from the beginning of chemotherapy was 116 days for the responders (partial remissions and non-total failures) and 110 days for the whole group of patients.

The most frequent side-effect was alopecia. It was almost complete and was observed in all patients who received more than one dose; it was reversible in those who were followed up for long enough (2 or 3 months after discontinuance of the drug).

Vomiting and/or nausea were present in 15 patients. This side-effect was not very troublesome with this intermittent schedule. One patient had diarrhea and another had an erythematous skin rash after the 4th dose. Stomatitis was observed in one single case, after the 3rd dose.

Clinically there were no signs of cardiac toxicity. However CPK was increased above the normal values in 4 cases and was accompanied by a first grade atrioventricular block in one case. No other ECG abnormalities were observed.

Hematologic toxicity was frequent but usually not severe and almost always reversible within 3 weeks. Leukopenia (less than 3000 leukocytes/mm^3) was observed in 21 among 23 evaluable cases; thrombocytopenia (less than 100000 platelets/mm^3) in 8 cases and anemia (hemoglobin less than 10 g/100 ml) in cases (Table 1). One patient died 6 days after the first dose adriamycin, with severe leukopenia (350 leukocytes) and a normal platelet level. The blood count was normal before treatment.

Table 1. Adriamycin in lung cancer. Hematologic toxicity

Leukopenia	< 3.000 WBC/mm³	21/23
	< 1.000	2/23
Thrombocytopenia	< 100.000 plat./mm³	8/23
	< 30.000	2/23
Anemia Hb	< 10 g/100ml	4/22
	decrease ≧ 20%	5/22

The tuberculin skin test was negative before treatment and remained negative in 3 cases. It was positive and remained positive in 11 cases. The skin test became negative during treatment in 3 cases. It was not repeated during treatment in the other patients.

Conclusions

This study establishes that an intermittent dose schedule of adriamycin (75 mg/m I. V. every 3 weeks) is able to produce partial regression in far-advanced lung cancer

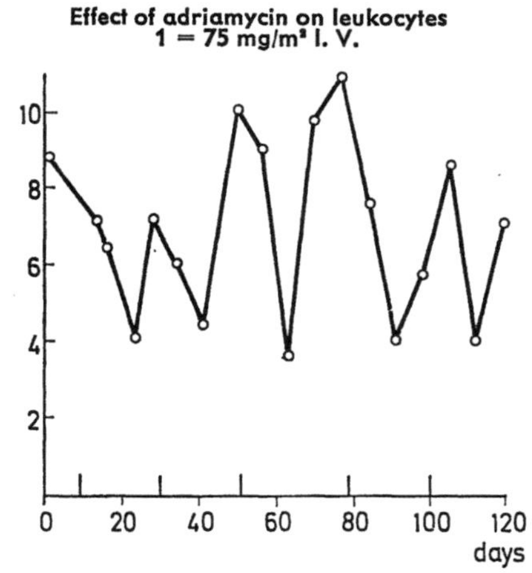

Fig. 1

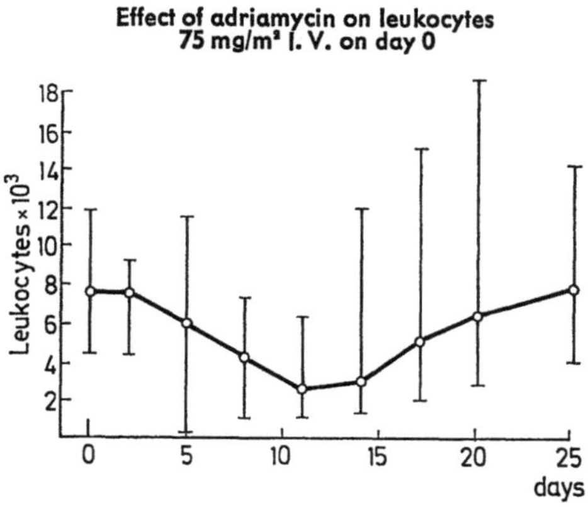

Fig. 2

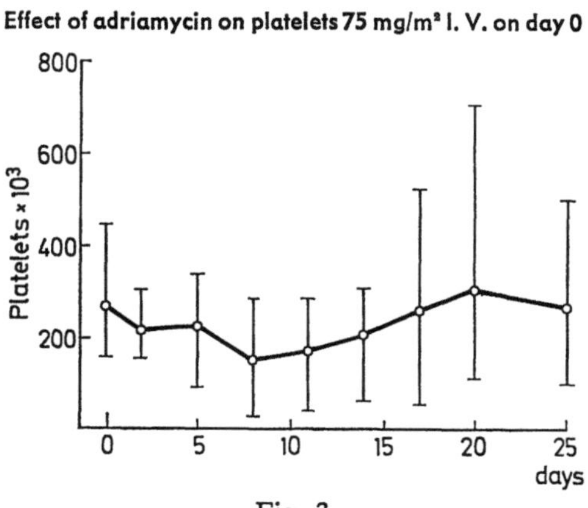

Fig. 3

Fig. 1. Effect of 5 doses of adriamycin on the leukocytes

Fig. 2. Median values of the leukocytes (and range) between 0 and 25 days after one dose of adriamycin

Fig. 3. Median values of the platelets (and range) between 0 and 25 days after one dose of adriamycin

The duration of the remission is not very long and the survival time of the patients does not seem markedly increased. An interval of 3 weeks between the injections seems adequate: in almost all cases leukocytes and platelets had returned to the normal pre-treatment value after 3 weeks (Figs. 1; 2; 3). The dose schedule was well tolerated. Cardiac toxicity was very limited and without clinical significance in this small group of patients. One peculiar feature of this regimen is the almost complete absence of stomatitis which was a frequent complication with other schedules. Comparing daunomycin and adriamycin, the latter seems much more effective in lung cancer, even at a dosage which produces less severe toxicity [4].

References

1. Groupe Coopérateur d'éssais Thérapeutiques sur les Cancers Broncho-Pulmonaires. Europ. J. Cancer **4**, 437-445 (1968).
2. Cooperative Group for Clinical Trials in Lung Cancer (E. O. R. T. C.). Europ. J. Cancer **7**, 477-478 (1971).
3. Di Fonzo, G., Gambetta, R., Bonadonna, G. Preliminary studies on the distribution and excretion of ^{3}H-adriamycin in man. First Congress of the European Association for Cancer Research, 1970 (abstract), p. 25.
4. Kenis, Y., Brule, G. Europ. J. Cancer **6**, 155-156 (1970).

Adriamycin in Adults with Neoplastic Diseases[1]

Irwin H. Krakoff

Medical Oncology Service, Department of Medicine, Memorial Hospital for Cancer and Allied diseases
and the
Division of Chemotherapy Research, Sloan-Kettering Institute for Cancer Research, New York, New York

It is perhaps inappropriate for my report to follow that of Dr. Bonadonna who has had by far the most extensive experience with adriamycin and whose experience has been substantially better than we have seen.

Patient Selection

Patients treated with adriamycin at Memorial Hospital were those with far-advanced non-resectable malignant neoplasms for whom no conventional therapy was considered likely to be useful. In most instances patients had received extensive prior radiotherapy and chemotherapy. Only patients whose disease could be objectively evaluated were considered for study with adriamycin. An exception to this was in the case of acute leukemia; since our current concepts of treatment for acute leukemia require intensive cyclic multiple drug therapy, adriamycin included in such regimens could not be evaluated as to its contribution to remission and survival. Adriamycin has been used in some children with acute leukemia as a single drug and these studies will be reported separately by Dr. Tan.

Dosage

Dr. Bonadonna has proposed four separate dosage schedules for adriamycin administration. In our studies we have used his schedule C almost universally, i.e., the daily injection of adriamycin for three consecutive days, then a four day rest period, then another three doses, another four day rest period and so on until bone marrow toxicity required the cessation of therapy. In an attempt to achieve more prolonged maintenance therapy, particularly in patients treated on an ambulatory basis, some patients were treated with a modification of his schedule A, i.e., three consecutive daily doses then a four day rest period and then one dose weekly. The doses used in both of these schedules were 0.3 or 0.4 mg/kg per dose. In our patient

1 Supported by Grants CA-08748 and CA-05826 from the National Cancer Institute.

population, using schedule C it was uncommon to get beyond the second week of adriamycin administration before bone marrow toxicity occurred. Using the modified schedule A it was often possible to continue therapy for several weeks.

Toxicity

Since a major criterion of the adequacy of drug trials was treatment to bone marrow depression it is obious that nearly 100% of patients adequately treated demonstrated bone marrow depression. However, irreversible or fatal bone marrow depression was not seen in this series.

An additional common manifestation of toxicity was stomatitis. This occurred as generalized mucosal erythema initially, progressing to oral ulceration. The stomatitis healed promptly with cessation of drug administration. In some debilitated patients stomatitis was complicated by secondary monilial infection and occasionally there was confusion of monilial infection with oral ulceration caused by adriamycin. Alopecia was very common in our series of patients treated with adriamycin (Table 1).

Table 1. Adriamycin toxicity — adult patients

	No. evaluable	Incidence	%
Leukopenia	49	40	82
Thrombocytopenia	48	29	60
Stomatitis	49	25	51
Alopecia	46	11	24

Although, as indicated above, this group of patients suffered from far-advanced cancer, in many cases with extensive pulmonary involvement, there were no episodes during these trials which could be interpreted as cardio-pulmonary toxicity. It should be noted however, that a contribution of adriamycin to cardio-pulmonary complications of cancer cannot be entirely eliminated. Adriamycin produces severe local inflamation and necrosis when extravasated from a vein. This was not a problem in our series since it was routinely injected through the tubing of a rapidly flowing intravenous infusion.

Therapeutic Results

Dr. TAN will report those portions of Memorial Hospital experience which we find encouraging. In adults at the present time, we have seen little response of various solid tumors to treatment with adriamycin. To date, 86 adults have been started in clinical trials with adriamycin. Since some of these are still in the midst of their trials as of August 1, 1971, only 50 can presently be evaluated (Table 2). Of those 50 patients, Category I responses (Karnofsky Criteria) have occurred in two patients with lymphosarcoma and one with a soft tissue sarcoma. Lesser responses have been seen in one patient with Hodgkin's disease, one with Ewing's sarcoma, one with colon cancer, one with kidney cancer and two with carcinoma of the bladder.

Table 2. Adriamycin — clinical effects — adults

Diagnosis	No. treated	No. adequate	Categories of response I-A I-B	II	O-C	O-B	O-A	O-O
Hodgkin's disease	8	6				1 (1/2)		5
LSA	2	1	1 (5+)					
Lymphoma cutis	1	1	1 (3+)					
Rhabdomyosarcoma	3	1	1 (4)					
Soft part sa's misc.	12	7						5
Neuroplastoma	2	1						1
Osteosarcomas	4	2			1			1
Head and neck	7	2						2
Lung	7	4						3
Bowel	7	5				1 (4)		4
Bladder	7	5		1 (4)			1 (1/2)	3
Prostate	2	2						2
Kidney	5	2				1 (1 1/4)		1
Wilms'	2	1						1
Breast	3	2						2
Cervix	5	3						3
Misc.	9	5						5
Total	86	50	3					

We believe it likely that the substantially better results seen in children treated at Memorial Hospital is a reflection of the somewhat different pattern of solid tumors which occur in children and although the number of cases is very small we are slightly encouraged by the responses in two patients with soft tissue sarcomas and one with Ewing's tumor.

Future Plans

We propose to continue our studies of adriamycin emphasizing tumors in those categories which have shown some evidence of responsiveness in our experience and those of others. The experience in acute leukemia although not evaluable from the standpoint of adriamycin alone suggest that it is a satisfactory component of multiple cyclic drug programs and in patients resistant to other kinds of therapy may itself produce brief remissions.

References

1. Arcamone, F., Franceschi, G., Tneco, S., Selva, A.: Tetrahedron Letters **13**, 1007-1010 (1969).
2. Di Marco, A., Gaetani, M., Scarpinato, B.: Cancer Chem. Rept. **53**, 33-37 (1969).
3. Bonadonna, G., Monfardini, S., De Lena, M., Fossati-Bellani, F.: Brit. med. J. 1969 III, 503-506.
4. Bonadonna, G., Monfardini, S., De Lena, M., Fossati-Bellani, F., Beretta, G.: Cancer Res. **30**, 2572 (1970).
5. Karnofsky, D. A., Burchenal, J. H.: The clinical evaluation of chemotherapeutic agents in cancer. In: MacLeod, C. M. (Ed.): Evaluation of Chemotherapeutic Agents. New York: Columbia University Press 1949.

Adriamycin in the Treatment of Acute Leukemias

G. MATHÉ, J. L. AMIEL, M. HAYAT, F. de VASSAL, L. SCHWARZENBERG, M. SCHNEIDER, C. JASMIN, and C. ROSENFELD

Institut de Cancérologie et d'Immunogénétique, Hôpital Paul-Brousse, and Service d'Hématologie de l'Institut Gustave Roussy, 94-Villejuif, France

With 2 Figures

This paper describes the results of our clinical trial of adriamycin in the treatment of acute leukemias, done after the first results of Bonadonna et al. [1]. A trial of this drug in solid tumors is at present being made by the "clinical screening group" of the European Organisation for Research on the Treatment of cancer (E.O.R.T.C.).

Patients and Dose Schedules

Patients

There were 90 patients in the trial, 63 with acute lymphoid leukemia and 27 with acute myeloid leukemia. There were 31 females and 59 males, aged from seven months to 74 years. All the patients were in the perceptible phase of their disease, the lymphoid leukemias were in their 1st-5th visible phase, and the myeloid in their 1st-3rd phase.

Dose Schedules

Adriamycin was given intravenously 0.4 mg/kg/d for 4 days, with a 3-day interval between courses of injections. The number of courses of injections varied from one to four (total dose 40—250 mg).

Observation

The patients were all treated in hospital and examined clinically twice a day, and parti cular attention given to the cardiovascular system. Haematological examinations were made three times a week with an immunological "check-up" before and after the treatment. An electrocardiogram and estimation of the serum creatinine phosphokinase (CPK) were made before, during and after the chemotherapy.

Tolerance

67 of the 90 patients had severe bone marrow aplasia with polymorphonuclear leucocyte counts below 300/cmm and platelet counts below 50000/cmm. 20 patients

require to be nursed in pathogen-free rooms for a long time. One patient with complete absence of platelets and neutrophils died of a meningeal haemorrhage.

Four patients showed a good haematological tolerance; 19 showed some degree of hypoplasia (Table 1). We noted, in 20 cases, oral ulcerations, and alopecia in 40 cases (Table 1).

Table 1. Tolerance to adriamycin treatment (90 patients)

Severe bone marrow aplasia	Bone marrow hypoplasia	Good haematological tolerance	Buccal ulcerations	Alopecia	Cardiovascular toxicity	
					Death by acute heart failure	Cardiomegaly ECG and biochemical anomalies
67	19	4	20	40	1	3

Cardiovascular toxicity

The cardiovascular toxicity was very pronounced in one patient. A child who had been given two courses of 25 mg/d for 4 days developed acute heart failure 3 days after the end of the second course of treatment. There was intense dyspnoea, cyanosis, diminution of heart sounds and tachycardia of 140/min. A chest X-ray showed the heart was enlarged. Despite treatment, the child died after 3 days in a state of severe heart failure and circulary collapse.

In three cases cardiac toxicity was detected by cardiomegaly, ECG and biochemical tests (Table 1). The only sign of cardiac toxicity is often a rise in the mean CPK, initially the mean value was 18.3 μM/ml/min, and rose at the end of the first course to 30 μM/ml/min (a significant difference: $t = 3.41$; $p < 0.01$).

Immune Responses

The effects of adriamycin on the immune responses were studied in 36 patients; the results are summarized in Table 2. 17 patients were positive before treatment, 10 remained positive after; 19 patients were negative before treatment, one became positive.

Table 2. Effects of treatment on immune responses

	Negative remained negative	Positive remained positive	Negative became positive	Positive became negative
Acute lymphoid leukemia	16	7	0	5
Acute myeloid leukemia	2	3	1	2

Results

Acute Lymphoid Leukemias

Table 3 shows that 24 apparently complete remissions and seven incomplete remissions were obtained in 63 patients; these remissions occurred in both the first visible phase of the disease and in subsequent relapses. The remissions were generally induced, in 2 or 3 weeks, or even within one week (Fig. 1).

Table 3. Results obtained in acute lymphoid leukemia with adriamycin (63 patients)

Phase	Remissions		Failures	
	Apparently complete	Incomplete	Partial	Total
1	5	1	1	5
2	10	4	1	7
3	4	2	3	7
4	3			5
5	2			3
Total	24	7	5	27

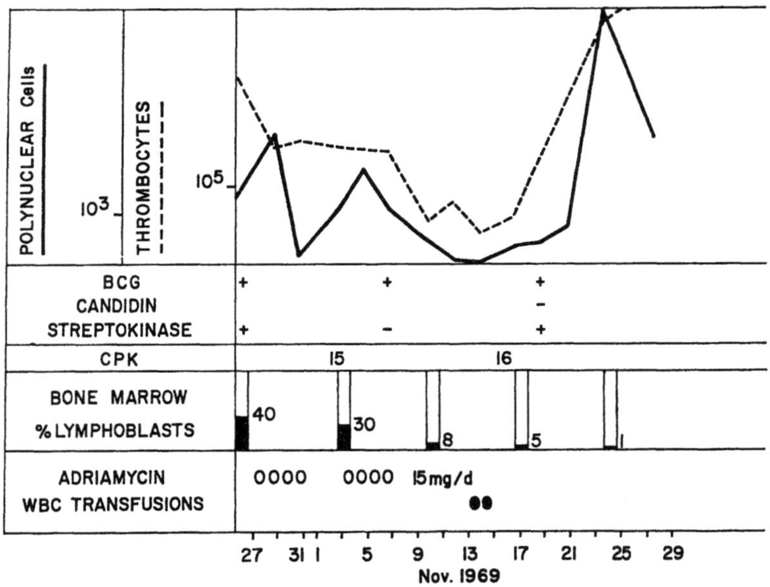

Fig. 1. Remission obtained by adriamycin in a case of acute lymphoid leukemia

Acute Myeloid Leukemias

Table 4 shows that 8 apparently complete remissions and 3 incomplete remissions were obtained in the 27 patients treated. The complete remissions were seen in patients in the first, second or third visible phase of the disease. Rapidity of onset was similar to that in acute lymphoid leukemia (Fig. 2).

Table 4. Results obtained in acute myeloid leukemia with adriamycin (27 patients)

Phase	Remissions		Failures	
	Apparently complete	Incomplete	Partial	Total
1	5	2	2	7
2	2	1	1	3
3	1			3
Total	8	3	3	13

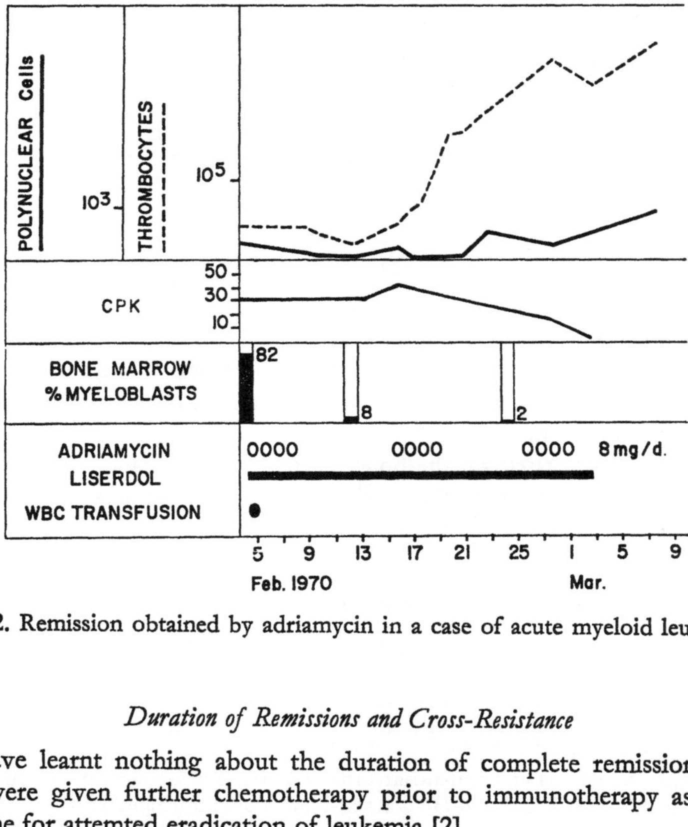

Fig. 2. Remission obtained by adriamycin in a case of acute myeloid leukemia

Duration of Remissions and Cross-Resistance

We have learnt nothing about the duration of complete remissions as these patients were given further chemotherapy prior to immunotherapy as part of a programme for attemted eradication of leukemia [2].

The remissions obtained in patients who were resistant to the combination of prednisone and vincristine have confirmed that there is no cross-resistance between these drugs and adriamycin. Four patients who were resistant to adriamycin were later found to be resistant to daunomycin (3 total failures, 1 partial failure).

Discussion and Conclusions

This trial has confirmed the effectiveness of adriamycin for the treatment of acute leukemias. The value of this drug is twofold: 1) it enabled a considerable percentage (24 out of 63 = 38%) of remissions to be achieved in acute lymphoid leukemias that were resistant to the combination of vincristine and prednisone; 2) it caused remissions in a noticeable percentage of patients with acute myeloid leukemia (8 complete remissions in 27 patients = 30%). The drawback to adriamycin is its high toxicity causing marrow aplasia, mucosal ulceration and cardiac failure (one death in 90). This trial has not established the place of adriamycin in the treatment of acute leukemias, but it has given some indications, in particular, the comparison of adriamycin and daunomycin under the same conditions, i.e. in acute lymphoid leukemia resistant to vincristine and prednisone as well as in acute myeloid leukemia.

In effectiveness, adriamycin compares favourably with daunomycin which gives remission in 25% of patients with acute lymphoid leukemias, not selected for their drug resistance, and 26% incidence of remission in acute myeloid leukemias [3].

The toxic effects on the haematopoietic system and mucosae seem to be of the same order as with daunomycin. Too few cases have been studied to assess the relative effects of these two drugs on the heart; in 112 patients given daunomycin [3] we have seen 15 major abnormalities of cardiac function (13.4%) with five deaths (4.5%), and in 90 patients given adriamycin there were four cases with severe cardiac lesions (4.3%) and one death (1.1%).

Summary

Adriamycin was given to 90 patients with acute leukemia, during the perceptible phase of the disease.

In 63 cases of lymphoid leukemia, 24 complete and 7 incomplete remissions were obtained. In 27 cases of acute myeloid leukemia, 8 complete and 3 incomplete remissions were obtained.

The drug was very toxic on the bone marrow, which renders necessary the use of pathogen-free rooms; its cardiac toxicity is definite but less frequent and less severe than the toxicity of daunomycin.

References

1. Bonadonna, G., Monfardini, S., De Lena, M., Fossati-Bellani, F.: Brit. med. Ass. 3, 503 (1969).
2. Mathe, G., Amiel, J. L., Schwarzenberg, L., Schneider, M., Cattan, A., Hayat, M., De Vassal, F., Schlumberger, J. R.: In: Recent Results in Cancer Research, Vol. 30, Berlin-Heidelberg-New York: Springer 1970.
3. E.O.R.T.C.: "Leukemia and hematosarcoma" cooperative Group. In: Recent Results in Cancer Research Vol. 30, Berlin-Heidelberg-New York: Springer 1970.

Cooperative Clinical Study on Adriamycin in Advanced Lung Tumors

C. Praga

Istituto di Clinica Medica dell'Universita' di Milano
(Direttore: Prof. E. Polli)

With 4 Figures

Experimentally a better therapeutic index has been found for adriamycin than daunomycin [4] and in preliminary clinical trials [1] adriamycin proved to be more active than daunomycin in solid tumors in general, and in lung tumors in particular [2].

Therefore the Clinical Pharmacology Group of the Lombardy Region (G.F.C.R.L.)[1] headed by Prof. Polli has planned a cooperative study on the effect of adriamycin in advanced inoperable lung tumors which was activated at the beginning of 1970.

50 patients were admitted to a first protocol with the dosage schedule shown in Table 1. Since no data were available concerning long-term treatment with adriamycin,

Table 1. Adriamycin in advanced lung tumor

	Dosage schedule
First protocol	0.5 mg/kg/d I.V. for 3 consecutive days After 7—10 day interval: three days course at the same dosage After 7—10 day interval: 0.5 mg/kg/d every other day for 3 days
Second protocol	1 mg/kg/d for 2 consecutive days After 21 day interval: 0.5 mg/kg/d for 2 consecutive days If objective regression is obtained, repeat the course every 3 weeks. Dosage adjusted according leukocyte and platelet levels.

1 G.F.C.R.L. (President: Prof. E. Polli; Secretary: Prof. C. Praga).
Participants in the trial:
Prof. Polli (Istituto di Clinica Medica dell'Università — Milano); Prof. Pasargiklian (Clinica Tisiologica dell'Università — Milano); Prof. Taglioretti (Ospedale di Angera); Prof. Ratti (Ospedale di Busto Arsizio); Prof. Della Porta (Ospedale di Casorate Primo); Prof. Pilotti (Ospedale di Mariano Comense); Prof. Chirico (Ospedale di Merate); Prof. Silvestrini (Ospedale di Niguarda-Carati); Prof. Nocca (Ospedale di Nogara); Prof. Castelfranco (Ospedale di Palazzolo sull'Oglio); Prof. Confalonieri (Ospedale di Rho); Prof. Melloni (Ospedale di Romano Lombardo); Prof. Mascaretti (Ospedale di Sarnico); Prof. Franzini (Ospedale di Saronno); Prof. Riva (Ospedale di Vimercate).

the evaluation of the results was carried out according to the KARNOFSKY's classification [5] one month after the end of treatment. Patients then left the trial and the various investigators were allowed to carry on with successive treatments according to their own judgement.

More recently, a new trial has been started with a dosage schedule (Table 1) fixed according to experience gained with the first protocol. This report is mainly concerned with the results obtained with the first treatment schedule; however, the preliminary results obtained with the second one are also discussed, as the comparison between them is of interest with regards to the toxicity of the drug.

Patients

The characteristics of the patient group treated with the first protocol are shown in Table 2: 41 out of 50 patients admitted to the study received an adequate treatment,

Table 2. Adriamycin in advanced lung tumor

Treated patients	50
Adequate treatment	41
Median age	60 (35—73)
Female/male	4/37
Previous treatment:	
None	11
Radio- and/or chemotherapy	30
Diagnosis to adriamycin therapy:	
Median interval (weeks)	22

that is, at least 2 complete courses equal to 3 mg/kg of the drug. No age limit was placed on the admission to the study, but subjects with previous myocardial lesions or with defective cardiac function were not accepted. Most patients (30/41) had already been treated with other chemo- and/or radio-therapy and 22 weeks was the median interval between the diagnosis of a lung tumor and the beginning of treatment with adriamycin. Patients were examined by chest radiography and stratigraphy, bronchoscopy and biopsy. The histology of all the cases could not be obtained; however, treatment was carried out only when the diagnosis of a lung tumor was certain.

Results

9 out of 41 patients are included in Category I (Table 3). A fairly large number of cases (Category 0-A: 11/41) showed only subjective improvement, i.e. reduction of pain, coughing, dyspnea, fever and improvement of physical state and of appetite. Other patients experienced only objective improvement (Category O-B: 4/41) or both objective and subjective improvement (Category O-C: 4/41) lasting less than one month. The cases classified as Category I (Table 4) were: oat-cell carcinoma 1, squamous-cell carcinoma 5, anaplastic carcinoma 2. and one case in which the histological diagnosis could not be obtained. Two cases (P.D. and F.S.) have received further courses of adriamycin besides those established by the protocol, and the

Table 3. Adriamycin in advanced lung tumor evaluable cases: 41 results (Karnofsky's categories)

Category O				I		
O	A	B	C	A	B	C
13	11	4	4	6	3	—

Table 4. Category I cases

Patient	Age	Sex	Adriamycin mg/kg	Hystological type	Survival (weeks) from diagnosis	treatment
P.D.	42	M	12	Oat cell cancer	90	88
L.O.	62	M	4.5	Squamous cell cancer	96[a]	72[a]
O.G.	43	M	4.5	Squamous cell cancer	60	32
B.R.	61	M	4.5	Squamous cell cancer	88[a]	52[a]
C.V.	44	M	3.5	Squamous cell cancer	52	26
F.S.	68	M	5	Squamous cell cancer	112[a]	48[a]
F.C.	69	M	4.5	Anaplastic cancer	64	44
G.G.	68	M	3	Anaplastic cancer	56	40
G.R.	49	F	4.5	Unknown	48	28

[a] Patient still alive.

12 mg/kg received by patient P.D. represents the highest total dose administered in this trial. Because of the relatively long survival of the patients classified as "Category I", the survival curves of the responsive and nonresponsive cases were compared (Fig. 1). The median survival of Category I cases (43 weeks) is almost double than

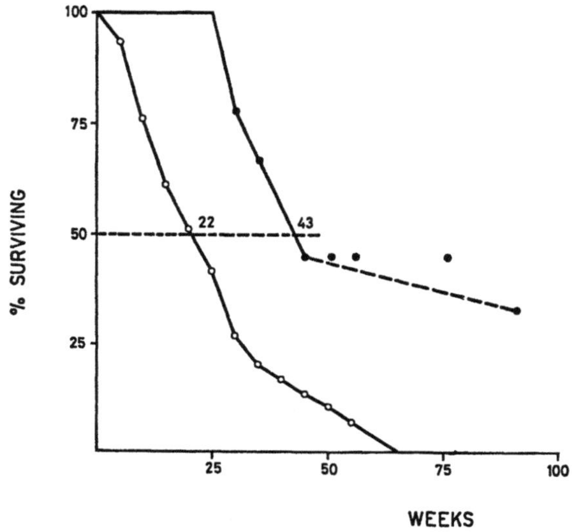

Fig. 1. Survival from adriamycin therapy. ○ "Category 0" patients (29), ● "Category I" patients (9)

that of Category O cases (22 weeks). Also considering the survival from the diagnosis, the difference is clear (Fig. 2), rising from 47 to 64 weeks; in addition the curves tend to deviate more towards the lower part of the graph due to the presence of longterm survivors.

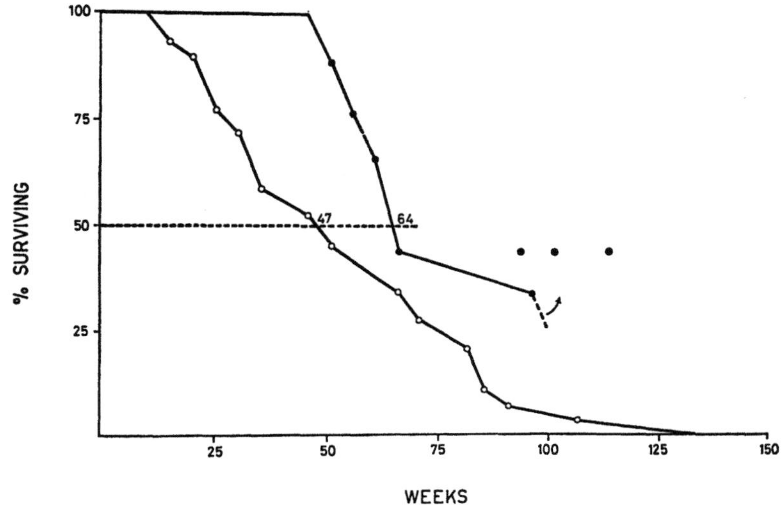

Fig. 2. Survival from diagnosis. ○ "Category 0" patients (29), ● "Category I" patients (9)

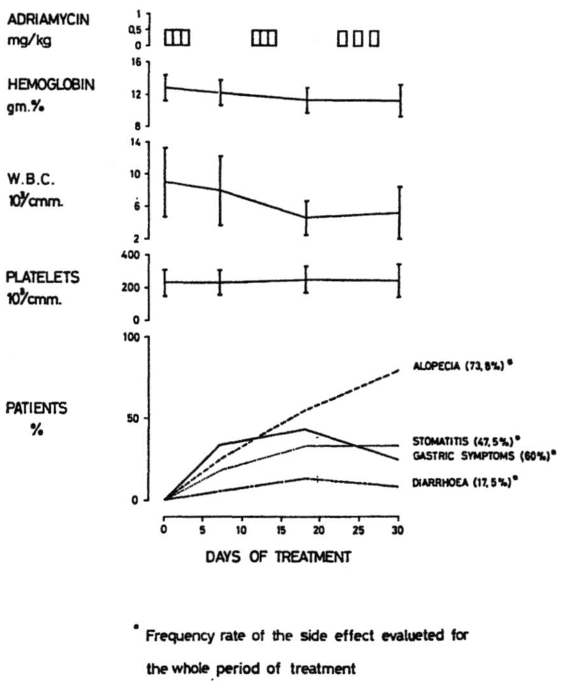

Fig. 3. Pattern of the hematological and clinical toxicity observed in the group of patients treated with the first dosage-schedule

Toxicity

Fig. 3 shows the frequency of the signs and symptoms of toxicity evaluated after every course. The greatest degree of bone marrow depression was observed after the second course with a remarkable decrease of leucocytes and less intense anemia; on the other hand, the platelets have not shown any significant variation throughout the treatment.

Nausea, vomiting and anorexia were particularly frequent after the first and the second courses, but clearly less after the thirdly one. Stomatitis and diarrhoea appeared more frequently after the second course and were closely connected with the drop in leucocytes. Alopecia appeared later and increased with duration of treatment.

The other parameters taken into consideration (Table 5) in order to evaluate the function of the various organs (blood sugar, B.U.N., serum uric acid, calcemia, transaminases, total proteins, albumin) have shown no significant change throughout the treatment. All the patients were followed up clinically and by serial ECG's for possible cardiac toxicity. A patient without a past history of cardiovascular disease died after the thirdly adriamycin cycle of unexplained acute heart failure. A week before death he presented with thrombophlebitis of the "vena axillaris" and was treated with heparin. Autopsy did not show any pathological findings at the level of the myocardium; however, a modest pericardial haemorrhagic effusion was present (ca. 700 ml) which could not have by itself caused the sudden heart failure. Two more patients showed reversible changes of the repolarization phase and a fourth patient dresented with atrial fibrillation only during the second course of treatment.

Table 5

	Basal	first cicle	second cicle	thirdly cicle
Blood sugar (mg %)	83.1 ±19.8	80.9 ±16.5	79.8 ±17.9	83.6 ±20.8
B.U.N. (mg %)	28.9 ± 8.2	31.4 ± 8.2	29.4 ± 7.8	29.0 ± 7.6
Uric acid (mg %)	4.39 ± 1.15	4.47 ± 1.13	4.67 ± 1.28	4.72 ± 1.31
Calcium (mg %)	9.75 ± 1.23	9.74 ± 0.98	9.95 ± 1.46	9.27 ± 0.98
S.G.O.T. (U/ml)	16.2 ± 8.7	19.9 ±15.3	22.1 ±19.8	22.6 ±14.2
S.G.P.T. (U/ml)	14.6 ±17.1	14.5 ±16.2	13.9 ±20.5	16.0 ±21.7
Total proteins (g%)	6.74 ± 0.54	6.50 ± 0.53	6.32 ± 0.56	6.43 ± 0.59
Albumin (g %)	3.22 ± 0.58	3.02 ± 0.50	2.93 ± 0.50	2.99 ± 0.55

In conclusion, out of 41 patients adequately treated, in one case adriamycin seemed responsible for a fatal cardiac toxicity. The overall toxicity data from this study, indicates that the second course, following the first one after an interval of 7—10 days, is poorly tolerated. With gastric symptoms of short duration and which are easily prevented, no acute toxicity of the drug was observed.

Based on these considerations, a new trial with the second dosage schedule shown in Table 1 has been started and 8 patients have so far been treated; 3 of them have already had an objective regression lasting more than one month.

It is interesting to observe how the pattern of toxicity associated with the treatment is clearly modified (Fig. 4). With the new protocol, the leucocyte drop previously observed does not appear and there is a clear rebound of leucocytes and platelets during the week preceeding the second course of treatment.

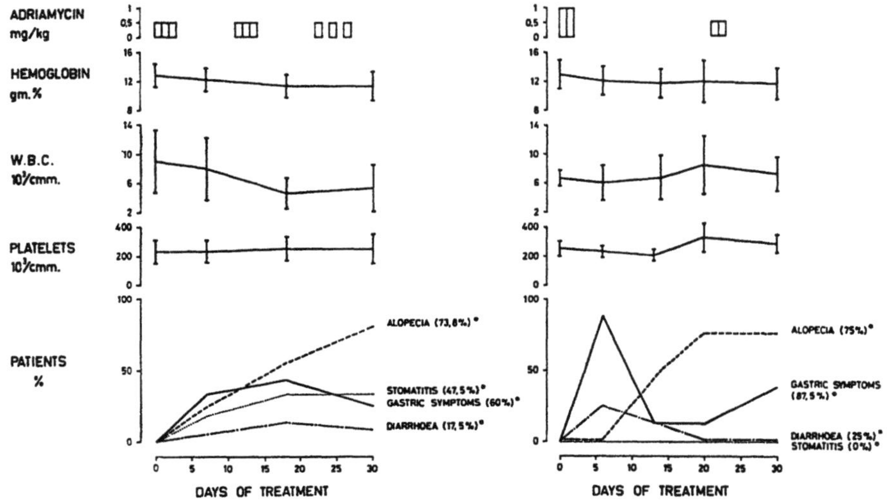

• Frequency rate of the side effect evalueted for the whole period of treatment

Fig. 4. Comparison between the pattern of the hematological and clinical toxicity observed in the groups of patients treated with two different dosage-schedules

The most promising result, however, is the complete absence of stomatitis, one of the most unpleasant symptoms with the first protocol.

After the first high-dose course the incidence of gastric symptoms, which seems clearly dose-dependent, is higher, but recedes fairly rapidly.

Comment

This study was planned in order to evaluate the percentage of objective results that could be obtained with adriamycin in patients with advanced lung cancer, already treated by conventional therapy. Owing to the difficulty of quantitative evaluation of lung tumor regression, the effect of adriamycin on the survival of treated patients has also been evaluated. Particular attention has been paid to the type of toxicity related to the dosage schedule.

Among the cases treated with the first protocol, adriamycin has proved to be capable of producing favourable objective changes for more than one month in different histological types of advanced lung tumors. The frequency of Category I responsive cases is 22% and their median survival, both from diagnosis and from adriamycin therapy, is greater than that of non-responsive cases.

A comparison with the results obtained with other drugs used in treating advanced lung cancer is not possible since, in non-randomized series, too many variables can interfere either in objective tumor regression or in patient survival. These results, however, justify a comparative randomized trial in the near future. The optimum posology has still however to be ascertained.

In our experience the first dosage schedule is rather toxic, especially the incidence of severe stomatitis, which may be complicated, in some cases, by infectious processes in the upper digestive and respiratory tracts. The preliminary results (8 cases in all) obtained with the second protocol suggest that the incidence and severity of the stomatitis may be considerably reduced by providing a longer interval (3 weeks) between the courses. It should be stressed that the high initial doses (1 mg/kg) are well tolerated except for some gastro-enteric disturbances which are frequent but easily overcome by conventional therapy. Whether this better tolerated treatment schedule will result in an equal or even better antitumoral effect, is now under investigation. Some experimental data on the distribution of adriamycin in the body [3] show a rapid uptake of adriamycin by the tissues and a slow release. The speed of penetration into the tissues is a function of the extracellular concentration, while the intensity of the "cell kill" produced by the drug is a function of the persistence of a high endocellular concentration over a long period of time [6].

The administration of high doses in a short time may achieve this goal, while a longer interval between the courses permits recovery of the normal tissues and reduces the toxicity of successive administrations.

References

1. Bonadonna, G., Monfardini, S., De Lena, M., Fossati-Bellani, F.: Brit. med. J. **1969 III**, 503—506.
2. Bonadonna, G., De Lena, M., Beretta, G.: Europ. J. Cancer **7**, 365—367 (1971).
3. Di Fronzo, G., Gambetta, R. A., Bonadonna, G.: Preliminary studies on the distribution and excretion of tritiated adriamycin in man. Comm. Europ. Meet. and 6 Meet. Therapeutic Chemistry, Bruxelles, 14—17 Sept. 1970.
4. Di Marco, A., Gaetani, M., Scarpinato, B.: Adriamycin (NSC — 123, 127): a new antibiotic with antitumor activity. Cancer Chemother. Rep. **53**, 33—37 (1969).
5. Karnofsky, D. A.: Clin. Pharm. Ther. **2**, 709—712 (1961).
6. Yesair, D.: Personal communication.

Adriamycin and Daunomycin in the Treatment of Vesical and Prostatic Neoplasias. Preliminary Results

M. Pavone-Macaluso and G. Caramia

Clinica Urologica dell'Universita' di Palermo (Direttore: Prof. S. Furnari)

Our preliminary experience with adriamycin in tumors of the bladder and prostate is of 25 treatments performed in 21 patients. The discrepancy between the numbers of patients and treatments is due to the fact that some patients were again submitted to various therapeutic cycles upon subsequent admissions. The treatments administered to the same patient at different times are considered individually, with regard either to a different schedule and way of administration of the drug, or to the interposition of other therapeutic means in the intervals between treatments.

In addition, 13 patients were treated with daunomycin, as shown in Table 1.

Table 1. General distribution of treatments with adriamycin and daunomycin in patients with vesical or prostatic tumors

Organ	Drug	Route of administration (N. of cases)				
		A. Intravenous	B. Intravesical			Total A+B
			a) Therapy	b) Prophylaxis	Total a+b	
1. Bladder	Adriamycin	8	4	3	7	15
	Daunomycin	0	4	7	11	11
2. Prostate	Adriamycin	10	—	—	—	10
	Daunomycin	2	—	—	—	2
Total 1+2	Adriamycin	18	4	3	7	25[a]
	Daunomycin	2	4	7	11	13
	Total (adriamycin daunomycin)	20	8	10	18	38

[a] 2 not controlled.

We shall refer only briefly to the use of daunomycin in urology, in view of the fact that this symposium is devoted chiefly to adriamycin. Some preliminary data on our experience with daunomycin have already been reported [9, 14, 15, 16].

Adriamycin was employed in 15 treatments of patients with bladder tumors, as well as in 10 cases of estrogen-resistant prostatic carcinomas. All the patients with prostatic cancer were treated intravenously. For bladder tumors, the intravenous route was employed in 8 cases of deeply infiltrating carcinoma (stage T 3 — T 4). In the remaining cases, where intravesical treatment was adopted, the tumor was confined to the most superficial layers of the bladder wall (stage Tx — T 2).

Papillary transitional cell carcinoma was the histologic diagnosis in all such cases. The topical treatment was employed both therapeutically (group A: 4 cases) and prophylactically (group B: 3 cases), the latter following transurethral surgery. Intravesical instillations are performed as follows: therapeutic treatments consist of 40 mg dissolved in 30 ml of sterile water to be retained for one hour, whereas for prophylactic use 10-20 mg is considered adequate.

The instillations are repeated at weekly intervals, but more frequent treatments my be given if deemed necessary.

The intravesical administration of adriamycin is well tolerated as regards both local and systemic effects. Mild and transient symptoms of cystitis were present in only 2 out of 7 cases. In no case was any change in the number of white cells or platelets observed. There were no other signs of systemic toxicity.

Attempts to recover adriamycin from the circulating blood after intravesical instillation of high doses have failed, showing that reabsorption of the drug through the intact bladder mucosa is minimal or absent.

The following clinical results were obtained after topical administration of adriamycin: an objective regression of the tumor was observed in 3 out of 4 treatments. In one case such a regression exceeded 50%; this patient was an 81-year-old woman with a papillary transitional cell carcinoma which had spread to the whole bladder mucosa, with severe anemia, profuse hematuria and renal insufficiency with hyperazotemia from urinary stasis in both ureters. Old age and poor general condition contraindicated total cystectomy, whilst a useful transuretral resection was impossible due to the bulk and the extension of the tumor. Prolonged topical treatment with adriamycin to a total dose of 500 mg was followed by a marked reduction in the size of the growth, accompanied by almost complete cessation of hematuria. Red cell counts and blood BUN reverted to normal. Residual tumor is still present and the left ureter is still dilated, but the patient has become capable of leading a normal life and enjoys reasonably good health.

No recurrences have so far occurred in patients treated prophylactically.

Daunomycin has also been well tolerated both locally and systemically. It has yielded comparable results if used intravesically for either a therapeutic or a prophylactic purpose.

Despite the limited number of cases and the short follow-ups, we consider the results rather encouraging so far. These drugs deserve, therefore, further trials in this field.

Adriamycin and daunomycin appear to be comparatively less active than thiotepa [6], peptidic sarcolysin complex [13] and other alkylating agents [15] in inducing tumor regression after intravesical instillations. On the other hand, this is counterbalanced by a lesser incidence of local inflammatory effects and the lack of any significant absorption into the systemic circulation that may lead to severe toxicity and even to lethal effects, as reported following thiotepa administration.

Topical use for therapeutic purposes may be indicated in widespread or diffuse non-infiltrating papillary bladder tumors.

Prophylactic administration appears particularly useful, if we consider that both adriamycin and daunomycin have displayed an activity that appears at least as good as that of other more dangerous drugs, notwithstanding the lack of significant local irritant properties and of reabsorption.

Only one recurrence was observed in a group of 7 patients treated prophylactically with daunomycin; it is noteworthy that such a recurrence occurred only after the treatment had been discontinued. The number of patients submitted to periodic intravesical instillations of adriamycin for prevention of recurrence of papillary bladder tumors is still small, but no tumors were found at cystoscopy in any of the 3 patients examined. The number of recurrent tumors in a comparable series of patients given no prophylactic treatment is as high as 60% [10, 12]. The number of recurrences is highest within the first six months following endoscopic or open surgery.

Our results were less encouraging in prostatic carcinoma, and especially in deeply infiltrating bladder cancer treated intravenously. The standard dose was 0.5 mg/kg for three consecutive days with intervals of ten days. However, we used a more prudent dose schedule when dealing with elderly, poor-risk patients, already treated by irradiation or with other antiproliferative agents, especially those suffering from prostatic carcinoma with widespread bony metastases. In the latter category of patients, a readjustment of the standard dose is necessary [16] since they are particularly liable to severe medullary depression.

Only 2 out of 7 patients with deeply infiltrating bladder carcinoma responded with marked symptomatic relief and in only one of them was an objective improvement demonstrated. This is at variance with the results obtained by Luce and Frey (personal communication), who observed marked objective regressions in 6 out of 11 treated patients. This discrepancy may be due to a difference in stage, grade, previous treatments and general condition of the patients or to the fact that Luce and Frey used higher doses than ours (and this may reflect again a difference in selection of patients). A 33-year-old patient with an anaplastic carcinoma displayed a dramatic subjective improvement after 3 cycles of adriamycin.

The subjective response consisted of total disappearance of dysuria, sciatalgia and hematuria, recovery of appetite and an increase in body weight of over 5 kg. An objective improvement was also apparent. A decrease of palpable tumor mass was felt by bimanual pelvic palpation and evaluated as superior to 50%. This was accompanied by the functional recovery of the left kidney that displayed no visible excretion of contrast medium prior to the treatment, due to infiltration of the pelvic ureter.

The extent of objective regression was sufficient to allow subsequent radical surgery (total cystoprostatovesiculectomy) to be performed with ease, whereas the tumor was initially considered as unresectable.

In estrogen-resistant prostatic cancers, subjective improvements were reported by the patients in 7 out of 10 cases. An objective reduction of palpable tumor mass was

Table 2. Bladder tumors treated intravenously with adriamycin

No	SEX	Tumor Grade	Stage	Cell type	Total dose (mg)	Side effects	Results Subjective	Objective	Survival months	Karnofsky-class
1	M	III	T 3	Trans.	220	Alopecia, vomiting, diarrhea.	Worse	Progression	6	O-O
2	M	II	T 4	Trans.	140	Alopecia	Disappearance of pain.	No	6	O-A
3	M	—	T 4,N,M	—	40	No	No	Progression	Dead	O-O
4	M	—	T 4, M	—	30	No	No	No	Lost to follow-up	
5	M	—	T 4, M	—	270	Alopecia, Stomatitis	No	No	Dead	O-O
6	M	II	T 3	Trans.	60	No	No	No	8	O-O
7	M	III	T 3	Anapl.	60	No	Not examined			
8	M	II	T 3	Anapl.	300	Alopecia	Marked improvement.	Regression over 50%	6	I-B

Table 3. Bladder tumors treated intravesically with adriamycin

No	Sex	Tumor Grade	Tumor Stage	Tumor Cell type	Way of administration	Total dose (mg)	Side effects	Results Subjective	Results Objective	Survival months	Karnofsky-class
					Group A) Therapeutic treatments						
1	M	II	T 1	Trans.	20 mg dissolved in 1.000 ml of 5% dextrose; daily irrigation for 3 days.	60	No	Improvement of general condition	Regression over 50%	8	O-B
2	F	II	T 2	Trans.	Weekly instillation of mg 40 for 3 weeks.	120	No	Improvement	Regression under 50%	1	I-A
3	F	II	T 2	Trans.	Weekly instillation of mg 40 for 2, then 20 mg for 15 weeks.	380	Irritative cystitis.	Improvement	Regression under 50%	3	I-A
4	F	II	T 1	Trans.	Instillation of 10 mg every 3 days for 4 days	40	No	No	No	5	O-O
					Group B) Prophylactic Treatments						
1	M	II	T 1	Trans.	20 mg i.v. + 10 mg intravesically: once a week for 8 weeks.	160 + 80	Mild cystalgia.	Improvement	No recurrence	3	
2	M	II	T ×	Trans..	Daily instillation of 20 mg for 3 days	60	No	Improvement	No recurrence	5	
3	M	II	T 2	Trans.	Daily instillation of 10 mg for 3 days.	30	No	Improvement	No recurrence	3	

Table 4. Estrogen-resistant prostatic carcinomas treated with intravenously adriamycin

No	Tumor			Total dose (mg)	Side effects	Results		Survival months	Karnofsky-class
	Grade	Stage	Cell type			Subjective	Objective		
1	III	T 4, M	Adenocarc.	80	No	Improvement of dysuria.	Decrease of sedimentation rate	1	O-C
2	III	T 4, M	Adenocarc.	60	Mild stomatitis	Improvement	Decrease of acid phosphatase	Dead at 7 mo.	O-C
3	IV	T 4, M	Adenocarc.	80	Subcutaneous petechiae	Improvement	Regression under 50%	Dead at 3 mo.	O-C
4	IV	T 4, M	Anaplastic	60	No	Improvement	Resumed am-lation and spontaneous micturition	3	O-A
5	IV	T 4, M	Anaplastic	60	No	Improvement	No	1	O-A
6	—	T 4, M	—	60	No	No	No	Dead at 15 days	O-O
7	III	T 4, M	Adenocarc.	100	Mild throm-bocytopenia	No	No	Dead at 1 mo.	O-O
8	II	T 3	Adenocarc.	180	Alopecia, mild leukopenia	Improvement	No	3	O-A
9	II	T 4, M	Adenocarc.	300	Alopecia	Improvement	Decrease of sedimentation rate	3	O-C
10	—	T 4	—	60	No	No	No	Not examined	

observed only once, although parameters indicative of an objective response, such as serum acid phosphatase levels and red cell sedimentation rate, were observed in 4 cases.

Furthermore, a striking, albeit transient, subjective remission was obtained in a patient suffering from a prostatic carcinoma metastatic to the dorsal spine with signs and symptoms of medullary compression including paresis of lower limbs.

Daunomycin administration was followed by almost complete disappearance of severe neurologic symptoms, while the patient resumed an almost normal gait.

The results seem, therefore, to be comparatively better in prostatic than in vesical cancers. In prostatic carcinoma the administration of adriamycin, despite lack of interference with the natural history of the disease, seems capable of obtaining some useful palliative effects, particularly after the stage of resistance to estrogens has been reached.

Toxicity was always moderate and reversible in our experience. Alopecia was a constant feature of intravenous therapy, whereas stomatitis and myelotoxicity were never of such severity as to cause interruption of the treatment.

Some details of the single treatments are given in the next part of our report. A comprehensive review of our data is given in Tables 2, 3 and 4.

As regards experience with other drugs in urologic tumors, the reader is referred to our previous work on this topic [1, 3—5, 7, 8, 11, 16]. The U.I.C.C. classification of tumors concerning the stage and modification of Broders' scale for histological grading and Karnofsky's classification of results were employed throughout, as indicated in our previous papers [2, 9, 16].

Summary

Intravesical instillations of adriamycin and of daunomycin are well tolerated by the bladder mucosa and do not give rise to systemic toxis effects. Topical administration of adriamycin has been followed by partial regression of the growths in 3 out of 4 treatments for vesical papillomatosis.

Prophylactic intravesical administration of adriamycin or daunomycin at weekly intervals following transurethral surgery has also proven of value. No recurrences have been observed in 3 patients treated with adriamycin, whereas only one recurrent tumor was found, after discontinuation of the treatment, in one out of 7 patients treated intravesically with adriamycin.

Our experience has yielded less favourable results in patients given adriamycin by the intravenous route. Only one objective regression, greater than 50%, accompanied by a marked subjective improvement, was obtained in a group of 8 patients with infiltrating carcinoma of the bladder. Symptomatic relief was also generally poor in this group.

The results were comparatively better in estrogen-resistant prostatic carcinoma. An objective regression of the tumor mass was present in only one out of 10 treatments, whereas a decrease in the levels of serum acid phosphatase and in red cell sedimentation rate was obtained in many cases. Subjective improvement was also relatively frequent, including one out of 2 patients treated with daunomycin.

References

1. Burt, F. B., Pavone-Macaluso, M., Horns, J. W., Kaufman, J. J.: J. Urol. (Baltimore) **95**, 51—57 (1966).
2. Furnari, S., Pavone-Macaluso, M., Lattuca, C., Verace, V., Longo, E.: Atti Soc. ital. Urol. **39**, 147—162 (1966).
3. Lattuca, C., Pavone-Macaluso, M., Verace, V.: Arch. Sic. med. Chir. **8**, 569—579 (1967).
4. Lattuca, C., Leone, G., Pavone-Macaluso, M.: Arch. Sic. med. Chir. **8**, 813—829 (1967).
5. Pavone-Macaluso, M., Leone, G., Lattuca, C., Palazzotto, G., Di Chiara, A.: I citostatici nel trattamento delle neoplasie dell'apparato urinario, p. 1—65. Palermo: Ediz. G. Denaro, 1967.
6. Furnari, S., Pavone-Macaluso, M., Lattuca, C., Leone, G.: Arch. ital. Urol. Nefrol. **41**, 187—198 (1968).
7. Pavone-Macaluso, M.: Atti soc. ital. urol. **41**, 413—422 (1968).
8. Pavone-Macaluso, M.: Verh. dtsch. Ges. Urol. 186—191 (1969).
9. Furnari, S., Pavone-Macaluso, M., Veroux, G., Mattea, E.: La chemioterapia antitumorale in Urologia. — Report 43th Meeting of the Italian Urological Association, Florence Nov. 5—8, 1970. Atti Soc. ital. Urol. **I**, sec. 2, 5—423.
10. Pavone-Macaluso, M., Furnari, S.: Chemoprophylaxis of papillary bladder tumours. Proc. XV Congr. Intern. Soc. Urol., Tokyo **2**, 377—383 (1970.)
11. Furnari, S., Pavone-Macaluso, M., Veroux, G., Mattea, E.: Urol. intern. (Basel) **26**, 311—318 (1971).
12. Pavone-Macaluso, M., Caramia, G., Rizzo, F. P.: The use of chemotherapeutic agents in the prophylaxis of papillary bladder tumors. Urol. Intern. (Basel) **26**, 302—310 (1971).
13. Pavone-Macaluso, M., Caramia, G.: Sull'impiego di un nuovo antiblastico, il complesso peptidico della sarcolisina (Peptichemio I.S.M.), nelle neoplasie dell'apparato urinario. Risultati preliminari. 1971 (In press).
14. Pavone-Macaluso, M.: Brit. J. Urol. **43**, 673—680 (1971).
15. Pavone-Macaluso, M.: Local chemotherapy of papillary neoplasias of the bladder. Panminerva Medica **13**, 240—249 (1971).
16. Pavone-Macaluso, M., Caramia, G., Vecchioni, M.: J. Urol. Néphrol. (Paris) 1971, (in press).

Clinical Trial of Adriamycin

C. ROZMAN*, E. SIMÓ CAMPS*, M. RIBAS MUNDÓ*, F. SOLSONA**, I. DANTART**, A. RAICHS***, and M. GIRALT***

* 2ª Cátedra de Medicina Interna. Departamento de Hematología. Barcelona.
** Departamento de Radioelectrología. Ciudad Sanitaria de la Seguridad Social. Zaragoza.
*** Departamento de Hematología. Ciudad Sanitaria de la Seguridad Social. Zaragoza/ Espana.

Adriamycin is a new antibiotic of the anthracycline group with a very important antitumor activity [1, 2]. In the recent literature several contributions to the clinical and therapeutic aspects of adriamycin in human tumors have been published [3—8].

We present in this paper the results of a cooperative clinical trial carried out in three Spanish institutions. Table 1 shows the numbers of patients treated in each institution.

Table 1. Clinical trial of adriamycin (April, 1st—June 30th, 1971)

2da Cátedra de Medicina Interna. Departamento de Hematología. Universidad de Barcelona	35 cases
Departamento de Radioelectrología, Ciudad Sanitaria de la Seguridad Social. Zaragoza	18 cases
Departamento de Hematología, Ciudad Sanitaria de la Seguridad Social Zaragoza	11 cases
Total	64 cases

Patients and Methods

From April 1st to June 30th, 1971, we treated 64 patients with adriamycin: solid tumors with metastasis 51 patients, acute leukemias 13 patients. The distribution of these patients according to their diagnosis and to age and sex is shown in Tables 2, 3 and 4.

The dosage adopted for solid tumors was that of BONADONNA et al. [5]; adriamycin 0.5 mg/kg was administered i.v. through the tubing of a running intravenous infusion consecutively during 3 days followed by 7 days rest. In another group of patients with solid tumors we administered the same dosage of adriamycin associated with prednisone 2.5 mg/kg i.v. during the 3 days of treatment in an attempt to reduce the toxicity of adriamycin and improve therapeutic results (Table 5).

Table 2. Distribution of the patients according to their diagnosis

I. Solid tumors with metastasis		
Bronchogenic carcinoma	10	51
Breast cancer	8	
Lymphoma	6	
Ovarian tumor	4	
Gastric cancer	3	
Hepatoma	3	
Cervix cancer	3	
Other tumors	14	
II. Acute leukemia		
Myeloblastic	8	13
Lymphoblastic	1	
Stem-cell	2	
Blastic crisis (CML)	1	
Erytholeukemia	1	

Table 3. Age and sex distribution. Solid tumors

Diagnosis	<20		20–29		30–39		40–49		50–59		60–69		>69		Total
	♂	♀	♂	♀	♂	♀	♂	♀	♂	♀	♂	♀	♂	♀	
Bronchogenic carcinoma	—	—	—	—	—	—	2	—	3	1	3	—	1	—	10
Breast cancer	—	—	—	—	—	—	—	3	—	4	—	1	—	—	8
Hodgkin's disease	—	—	—	2	—	—	1	—	—	1	—	—	—	—	4
Lymphosarcoma	1	—	—	—	—	—	—	—	1	—	—	—	—	—	2
Ovarian tumor	—	1	—	—	—	1	—	—	—	—	—	—	—	2	4
Gastric cancer	—	—	—	—	—	—	—	—	—	—	1	1	1	—	3
Hepatoma	—	—	—	—	—	—	—	—	1	1	—	1	—	—	3
Cervix cancer	—	—	—	—	—	—	—	2	—	1	—	—	—	—	3
Other tumors	—	—	—	—	—	—	2	—	3	3	1	2	1	2	14
Total	1	1	—	2	—	1	5	5	8	11	5	7	3	2	51

Table 4. Age and sex distribution. Acute leukemia

Diagnosis	<10		10–19		20–29		30–39		40–49		50–59		>59		Total
	♂	♀	♂	♀	♂	♀	♂	♀	♂	♀	♂	♀	♂	♀	
Myeloblastic	—	—	—	1	1	2	—	1	—	—	—	1	1	1	8
Lymphoblastic	—	1	—	—	—	—	—	—	—	—	—	—	—	—	1
Stem-cell	—	1	—	—	—	1	—	—	—	—	—	—	—	—	2
Blastic crisis (CML)	—	—	—	—	—	—	—	—	—	—	—	1	—	—	1
Erythroleukemia—	—	—	—	—	—	—	—	—	—	—	—	—	1	—	1
Total	—	2	—	1	1	3	—	1	—	—	—	2	2	1	13

Table 5. Dosage schedule

Diagnosis	Group	Cases	Drug	Route	mg/kg	Days	Courses
Solid tumors	A	30	Adriamycin	i.v.	0.5	× 3 rest 7 days	3 or more
	B	21	Adriamycin	i.v.	0.5	× 3 rest 7 days	3 or more
			+				
			Prednisone	i.v.	2.5	× 3 rest 7 days	3 or more
Acute leukemia		13	Adriamycin	i.v.	0.4	× 1 rest 9 days	3 or more
			+				
			Ara-C	i.v.	2.5	× 5 rest 5 days	3 or more

The patients with acute leukemia followed a protocol similar to that of Crowther et al. [9], substituting adriamycin for daunomycin. The dosage schedule was: adriamycin 0.4 mg/kg day 1 i.v. + ara-C 2.5 mg/kg i.v. days 1—5 in one hour infusion, with at least 5 days rest between courses (Table 5).

All patients admitted to the solid tumor group required a diagnosis confirmed by biopsy. In all cases the tumor had already produced metastasis. 17 patients had previously received cobalt therapy and 5 patients some form of chemotherapy but in no case had adriamycin or daunomycin been given. The patients with acute leukemia were all diagnosed by bone marrow aspiration. No patient was admitted with a previous myocardial infarct, heart failure or any important cardiac alteration.

Before the beginning of treatment all patients had a complete blood count, bone marrow aspiration, urinalysis, BUN, uricemia, glucemia, calcemia, alkali phosphatases, SGOT, SGPT, ESR, protein electrophoresis and ECG. A complete blood count was performed 3 times a week, chemical tests and ECG once a week.

No new course of treatment was administered until the WBC count was > 1500 and the thrombocyte count > 50000.

The therapeutic effects of adriamycin in the solid tumors group have been classified according to the categories of response proposed by Karnofsky [10]. In the acute leukemia group we followed the criteria suggested by the Midwest Cooperative Study Group [11].

Results

Tables 6, 7, 8 and 9 summarize the results obtained in 51 patients with solid tumors and 13 patients with acute leukemia.

Solid Tumors

The best results were obtained in bronchogenic carcinoma: 3 out of 10 patients were classified in category I-B with objective regression of more than 50% of all measurable neoplastic disease during one month or more, and with great subjective improvement; 2 other patients were classifed in category I-A with distinct subjective benefit and favourable objective improvement but less than 50% of measurable neoplastic disease during one month or more.

Table 6. Solid tumors with metastasis

Diagnosis	Cases	Category of response (Karnofsky method)					
		O—O	O—A	O—B	O—C	I—A	I—B
Bronchogenic cancer	10	2	2	—	1	2	3
Breast cancer	8	2	1	—	—	4	1
Hodgkin's disease	4	1	—	—	1	2	—
Lymphosarcoma	2	—	—	1	—	1	—
Ovarian tumor	4	1	1	—	—	—	2
Gastric cancer	3	1	2	—	—	—	—
Hepatoma	3	—	3	—	—	—	—
Cervix cancer	3	1	1	—	1	—	—
Total	37	8	10	1	3	9	6

Table 7. Solid tumors with metastasis

Diagnosis	Cases	Category of response (Karnofsky method)					
		O—O	O—A	O—B	O—C	I—A	I—B
Hypernephroma	2	2	—	—	—	—	—
Vagina sarcoma	1	1	—	—	—	—	—
Rectum cancer	1	1	—	—	—	—	—
Esophagus cancer	1	—	1	—	—	—	—
Disseminated carcinomatosis	1	—	1	—	—	—	—
Thyroid cancer	1	1	—	—	—	—	—
Mesothelioma	1	1	—	—	—	—	—
Pancreas cancer	1	—	—	—	—	1	—
Teratocarcinoma	1	1	—	—	—	—	—
Penis cancer	1	—	—	—	—	1	—
Larynx cancer	1	1	—	—	—	—	—
Endometrium cancer	1	1	—	—	—	—	—
Mucocele appendix	1	1	—	—	—	—	—
Total	14	10	2	—	—	2	—

In breast carcinoma we obtained 1 out of 8 cases in category I-B and 4 in category I-A. Two out of 4 ovarian tumors were also classified in category I-B.

Our results in cases of lymphosarcoma were less satisfactory, although all of them were resistant to previous chemotherapy and had been treated initially by radiotherapy. Category I-A was obtained in 2 out of 4 lymphogranuloma patients and in 1 out of 2 lymphosarcoma patients.

In the miscellaneous group of tumors the best results (Category I-A) were obtained in one case of pancreas carcinoma and another case of penis carcinoma.

Table 8. Solid tumors with metastasis

Group	Cases	Category of response (Karnofsky method)					
		O—O	O—A	O—B	O—C	I—A	I—B
A Without pred.	30	14	6	—	2	5	3
B With pred.	21	4	6	1	1	6	3
Total	51	18	12	1	3	11	6

When we compare the results of the group of 30 patients not receiving prednisone with the group of 21 patients receiving prednisone (Table 8), it is surprising that 14 out of 30 patients in the first group did not manifest any subjective improvement (Category O-O) but only 4 out of 21 in the second group were classified in this category. These results suggest a higher incidence of subjective improvement in the group treated with prednisone although, possibly due to the small number of patients, these figures are not statistically significant. We must also note that the patients had not been randomized.

Acute Leukemia

The results are summarized in Table 9. Complete remission was obtained in 2 out of 8 patients with acute myeloblastic leukemia, partial remission was obtained also in 2 and no response in 4 patients. Of these 8 patients 6 were first-treated and the 2 that did not respond were relapses resistant to other therapy.

Table 9. Acute leukemia

Diagnosis	Cases	C.R.	P.R.	N.R.
Myeloblastic	8	2	2	4
Lymphoblastic	1	—	—	1
Stem cell	2	1	—	1
Blastic crisis CML	1	—	1	—
Erythroleukemia	1	—	—	1
Total	13	3	3	7

One patient with lymphoblastic leukemia, resistant to vincristine, prednisone and cyclophosphamide, failed to respond to adriamycin and ara-C. Of 2 patients with stem-cell leukemia, one had complete remission and the other showed no response.

In one case of MCL with blastic crisis a partial remission was obtained after 3 courses of adriamycin and ara-C. One patient with erythroleukemia failed to respond.

Toxicity

Side effects are summarized in Table 10.

In the present series the most frequent side effects were alopecia occurring in 51% of the patients, diarrhea in 42%, stomatitis in 20% and nausea and vomiting in 7%. In no case had the treatment to be discontinued because of these side effects.

Table 10. Toxicity

a) Clinical			Cases	%
Alopecia			33	51
Diarrhea			27	42
Stomatitis			13	20
Nausea and vomiting			5	7
b) Hematologic	Mild	Serious	Total	%
Anemia	21	3	24	37
Leukopenia	20	5	25	39
Thrombopenia	12	2	14	21
c) Cardiac	Mild	Serious	Total	%
	11	1	12	18

The hematologic toxicity was considered serious when the Hb level was below 6 g, leukocyte count < 1500, or thrombocyte count < 50000. Serious anemia was observed in only 3 cases, serious leukopenia in 5 patients and serious thrombopenia in 2 cases.

Cardiac toxicity was transient in 11 patients and only detected by ECG changes. One patient 10 min after the second infusion of adriamycin developed protodiastolic gallop rythm and pronounced ECG changes which were reversible after a few hours.

The statistical analysis of side effects comparing the cases treated with and without prednisone did not prove to be significant.

Discussion

The results presented in this paper confirm the activity of adriamycin reported by other authors [3—8] in the treatment of solid tumors and acute leukemias. They are in agreement with the good results presented by Bonadonna et al. (5) especially in breast and lung Ca. Their acceptable results in G.I. tract carcinoma could not be confirmed in our trial but this might be explained by the smaller number of these patients in our series. A similar explanation might be given with respect to our less encouraging results in lymphoma; on the other hand, all 6 patients had been previously treated with radio- and chemotherapy and presented a very bad prognosis.

In acute myeloblastic leukemia the results are very close to the percentage of remissions we obtain with daunomycin and ara-C using the same schedule of treatment. Mathé [6] has reported similar results in a larger number of patients with AML, administering adriamycin alone in 4 day courses.

One case of CML with blastic crisis showed a partial remission which must be considered a good result knowing the poor prognosis of this situation.

The toxicity in our series has been somewhat lower than the toxicity reported by Bonadonna et al. [5]. It is possible that schedule D with at least 7 days rest between courses of 3 days represents a good improvement in attempting better tolerance of the drug with acceptable antitumor effect.

Summary

The results of a clinical trial of adriamycin in 64 patients are presented. Of 51 patients affected by solid tumors with metastasis, 16 showed objective response; a subjective response was obtained in 17 and no response in 18. The best results were observed in lung and breast carcinoma. Of 13 patients with acute leukemia, 3 had a complete remission, 3 a partial remission and 7 no response.

We can conclude that adriamycin is an active drug in neoplastic disorders which deserves further trials in acute leukemia and in solid tumors, particularly in lung and breast cancer.

References

1. Di Marco, A., Gaetani, M., Scarpinato, B.: Cancer Chem. Rep. **53**, 33—37 (1969).
2. Arcamone, F., Franceschi, G., Tecno, S., Selva, A.: Tetrahedron Letters **13**, 1007 to 1010 (1969). Pergamon Press.
3. Bonadonna, G., Monfardini, S., De Lena, M., Fossati-Bellani, F.: Brit. med. J. 1969 III, 503—506.
4. Monfardini, S., Bonadonna, G., Di Pietro, S., Guindani, A., Fossati Bellani, F., De Lena, M.: Tumori **55**, 197—216 (1969).
5. Bonadonna, G., Monfardini, S., Di Pietro, S., Beretta, G., De Lena, M., Fossati-Bellani, F.: Clinical Evaluation of Adriamycin. Presented at the X International Cancer Congress. Houston 22—29 May 1970.
6. Mathe, G., Amiel, J. L., De Vassal, F., Schwarzenberg, L., Schneider, M., Jasmin, Cl., Rosenfeld, Cl.: Presse méd. **78**, 1997—1999 (1970).
7. Tan, G., Wollner, N., King, O., Iland, D.: Proc. Amer. Ass. Cancer Res. 1970 II.
8. Obrecht, P., Westerhausen, M., Simon, A.: Dtsch. med. Wschr. **9**, 369—376 (1971).
9. Crowther, D., Bateman, C. J. T., Vartan, C. D., Whitehouse, J. M. A., Maldas, J. S., Hamilton Fairley, G., Scott, R. B.: Brit. med. J. 1970 IV, 513—517.
10. Karnofsky, D. A.: Clin. Pharm. Ther. **2**, 709—712 (1961).
11. Hewlett, J. S., Batlle, J. D., Jr., Bishop, R. C., Fawlerw, M., Schwartz, S. D., Hagen, P. S., Zonis, J.: Cancer Chemother. Rep. **42**, 25—28 (1964).

The Effect of Adriamycin on Solid Tumors and Acute Leukemia of Adults and Children[1]

Lucius F. Sinks*, Engracio Cortes**, Jaw J. Wang*, and James F. Holland**

From the Department of Pediatrics* and Medicine A**, Roswell Park Memorial Institute, 666 Elm Street, Buffalo, New York 14203

With 2 Figures

Materials and Methods

Eighty-six patients were included in this study. There were 44 patients with leukemia of whom 30 had acute lymphocytic leukemia (26 children and 4 adults), 14 had acute myelocytic leukemia. In addition, there were 20 children and 22 adults with a variety of disseminated solid tumors. All children with leukemia had advanced disease, and had received all conventional chemotherapeutic drugs including daunorubicin[1] in most cases.

In patients with solid tumors, a complete response was defined as the complete disappearance of subjective and objective signs of cancer, and a partial response as an incomplete decrease in tumor size but at least 50 percent. The patients with acute leukemia were evaluated in terms of the establishment of a bone marrow remission. A complete remission marrow specimen (M-1) was defined as a cellular specimen containing 5% or fewer lymphoblasts and 40% or fewer lymphocytes and blast cells in acute lymphocytic leukemia, and no more than 5 percent blast cells, promyelocytes, monocytoid cells or other elements which could not be classified as more mature normal elements in acute myelocytic leukemia. A partial remission marrow specimen (M-2) was a cellular sample containing 5.1—25% lymphoblasts or 40.1—70% lymphocytes plus lymphoblasts in acute lymphocytic leukemia. The duration of a complete response was the duration of a bone marrow remission.

In children the following dose schedules were used:

1. Adriamycin 17.5 mg/m^2 daily intravenously for 2 consecutive days. Following a 3-day rest period, a second 2-day treatment program was given. After the third such course, in the absence of drug toxicity, adriamycin was continued once weekly.

2. Adriamycin 17.5 mg/m^2 daily for 3 consecutive days, following a 4-day rest period, the 3-day treatment program was repeated. After the drug toxicity was cleared, adriamycin was continued once weekly.

1 The drug was furnished by Farmitalia Research Laboratories, Milan, Italy. The study was supported by U.S. Public Health Service Grant Nos. CA-07918, CA-05834, and CA-2599.

3. Because of severe myelosuppression occurring in some patients, the dose schedule for solid tumors was subsequently modified to 20 mg/m^2 daily for 4 consecutive days. Repeat the course after a 17-day rest period.

4. In children with leukemia, some received adriamycin 40 mg/m^2 weekly.

In adults, the dose schedule was similar to children, that is 20 mg/m^2 daily for a total dose of 120 mg/m^2 in 10—12 days. Subsequently, the dosage was also modified to 35 mg/m^2 daily for 3 consecutive days; repeating the course after a 17 day rest period.

In the absence of tumor regression or leukemia remission, i. e. when the blast cells in the bone marrow persisted or increased after recovering from hematological toxicity, the study was terminated.

Pre-treatment and follow-up studies included complete blood counts, bone marrow aspirations, liver function tests, BUN, serum uric acid, electrocardiogram, and roentgenographic studies. Supportive therapy with antibiotics, whole blood and blood components were given as indicated.

Results

Response to Therapy

The spectrum of tumor response to adriamycin therapy is summarized in Tables 1 and 2. Ten of 20 children and 9 of 22 adults with solid tumors responded to adriamycin therapy. Complete remission with disappearance of pulmonary metastasis occurred in 1 of 1 Wilm's tumor, 1 of 5 reticulum cell sarcomas and 1 of 13 osteogenic sarcomas. Partial remission with greater than 50% tumor regression also occurred in

Table 1. Adriamycin in children and adults with solid tumor and leukemia

Disease	No. of patients treated	Patients with partial response	Patients with complete response
Wilm's tumor	1		1
Neuroblastoma	8	5	
Ewing's sarcoma	3	2	
Rhabdomyosarcoma	1		
Osteogenic sarcoma	13	4	1
Reticulum cell sarcoma	5	3	1
Lymphosarcoma	1	1	
Bronchogenic carcinoma	6	3	
Malignant teratoma	2		
Soft tissue sarcoma	2		
Hepatoma	1		
Chondrosarcoma	2		
Malignant melanoma	1		
Acute leukemia			
ALL	30	1	5
AML	14		1
Total	90	19	9

Table 2. Detailed analysis of patients with solid tumors responding to therapy

Diagnosis	Age	Tumor site		Previous therapy		Chemotherapy	Response to adriamycin	
		Primary	Metastasis	Surgery	Radiotherapy		Degree	Duration (months)
Wilm's tumor	5	Kidney	Lung, liver, pelvis	Resection	Yes	Yes	Complete	3
Neuroblastoma	6	Thorax	Scalp, epidura, marrow, bone	Laminectomy	Yes	Yes	Partial	1
Neuroblastoma	8	Abdomen	Liver	Resection	Yes	Yes	Partial	2
Neuroblastoma	2	Abdomen	Scrotum, marrow mediastinum	Biopsy	Yes	Yes	Partial	2,5
Neuroblastoma	7	Abdomen	Pelvis, marrow	Resection	Yes	Yes	Partial	1,5
Neuroblastoma	1	Abdomen	Scalp, orbit, marrow, bone	Biopsy	No	No	Partial	1
Ewing's sarcoma	17	Spine	Lungs	Lanimectomy	Yes	Yes	Partial	3
Ewing's sarcoma	11	Ulnar	Lungs	Biopsy	Yes	Yes	Partial	3_+
Osteogenic sarcoma	13	Femur	Lungs	Amputation	No	No	Complete	14_+
Osteogenic sarcoma	17	Femur	Lungs	Amputation	No	No	Partial	12_+
Osteogenic sarcoma	21	Femur	Lungs	Amputation	No	No	Partial	6
Lymphosarcoma	20	Lymph nodes	Marrow	Biopsy	Yes	Yes	Partial	2
Reticulum cell sarcoma	37	Lymph nodes		Biopsy	Yes	Yes	Partial	10_+
Reticulum cell sarcoma	12	Thorax	Oral cavity	Resection	Yes	Yes	Partial	2
Reticulum cell sarcoma	61	Testicle	Liver, marrow, lung	Surgery	Yes	Yes	Complete	5
Reticulum cell sarcoma	58	Nasopharynx	Lungs, abdominal wall, spine	Biopsy, laminectomy	Yes	Yes	Partial	4_+
Bronchogenic carcinoma (adenocarcinoma)	74	Lung		None	No	No	Partial	4_+
Bronchogenic carcinoma (adenocarcinoma)	56	Lung		None	No	No	Partial	4_+
Bronchogenic carcinoma (anaplastic)	64	Lung		None	No	No	Partial	4_+
Osteogenic sarcoma	37	Humerus	Lungs	Disarticulation	No	No	Partial	4
Osteogenic sarcoma	23	Humerus	Lungs	Amputation	No	Yes	Partial	5_+

5 of 8 neuroblastomas, 2 of 3 Ewing's sarcomas, 3 of 6 bronchogenic sarcomas, 1 of 1 lymphosarcomas, 3 reticulum cell sarcomas and 4 osteogenic sarcomas. Remission lasted from 1 to 10 + months with a median of 3 months. Onset of response was evident usually between days 14—21.

Thirty patients (26 children and 4 adults) with acute lymphocytic leukemia were treated with adriamycin. Five patients achieved a complete remission of 8,2,1,0.3 and 0.25 months duration respectively. The former patient was maintained on 6-MP therapy. One child achieved a partial remission of 10 days duration. Thirten adults with acute myelocytic leukemia were treated with adriamycin. One patient achieved a complete remission lasting for 4 months. There was no apparent difference in response with different schedules of adriamycin administration, however, all responders experienced significant degrees of leukemia.

Toxicity

The major forms of drug toxicity observed are summarized in Table 3. The column entitled "Patient Evaluation" gives the number of patients who were considered suitable for determining the drug toxicity in question. Some patients were excluded owing to inadequate data, others because the toxicity in question predated the therapy, and others because we could not exclude the underlying disease as the cause of the problem.

Table 3. Adriamycin toxicity

Problem	Patients evaluable	Patients with toxicity problems
Vomiting	77	35
Oral ulceration	82	57
Alopecia	71	71
Chemical cellulitis	86	4
Leukopenia	81	66
Thrombocytopenia	73	53
Anemia	68	53
Electrocardiographic changes	60	4

The most significant toxicity of adriamycin was bone marrow depression. Leukopenia of less than 2000/mm^3 occured in 66 of 81 evaluable patients. The nadir of leukopenia in patients with solid tumor occured from day 13—24, whereas the nadir of leukopenia in patients with leukemia occurred from day 6 to day 20. Thrombocytopenia occurred in 53 of 73 evaluable patients. The platelet count nadir generally paralleled the leukocyte depression. Platelet transfusions were commonly used at the depth of the platelet depression to prevent bleeding. Anemia was common and often severe enough to require blood transfusion.

Bone marrow depression was common after the completion of a course of adriamycin. During this depressed period, the bone marrow consisted predominantly of lymphocytes with a marked decrease in the erythroid and myeloid series. With recovery, the erythroid series in the marrow generally returned first followed by the myeloid series. There are four deaths directly related to adriamycin induced myelosuppression; three died on sepsis and one died of hemorrhage before recovery from marrow aplasia.

Vomiting occurred in 60% of patients but usually disappeared within 48 hours. Erythema and ulceration of the buccal mucosa, tongue, and gums developed in two-thirds of the patients. ome patients complained of dysphagia, without oral lesion, suggestive of esophageal mucositis.

Extravasation of adriamycin in the ante-cubital space caused chemical cellulitis and severe induration followed by limitation of motion of the joints of 4 patients.

Alopecia was prominent, occurring on all 71 evaluable patients and persisting as long as 2—3 months. One child ceveloped a diffuse erythema with vesiculation limited to previously irradiated skin areas, consistent with radiation dermatitis.

Reversible electrocardiographic changes including multiple ventricular extrasystoles, PR prolongation, or ST depression occurred in four of 60 patients. No evidence of myocardial change was observed in three of these patients who had post-mortem examination. No significant liver dysfunction was observed as determined by serial serum determinations. Two children developed hypertension and azotemia, but microscopic examination of kidneys revealed evidences of radiation nephritis.

Case Reports

Case 1: J. L. A five year old boy was found to have an abdominal mass in September 1968. A diagnosis of Wilms' tumor was established in October 1968 on a laparotomy with excision of the tumor and the left kidney. Post-operatively, he was treated with radiotherapy, and chemotherapy including vincristine and dactinomycin. In December 1969, he had progressive disease with pelvic and hepatic masses, as well as pulmonary metastases despite continuous vincristine and dactinomycin therapy.

Adriamycin in a total dose of 105 mg/m^2 body surface was given on December 21, 1969 through January 10, 1970 resulting in complete regression of pelvic, hepatic, and pulmonary metastases. The patient's course was complicated with severe bone marrow depression and infection from which he recovered. Pulmonary metastases recurred 3 months later while he was on adriamycin maintenance therapy.

Comment: In this patient, adriamycin caused clinical disappearance of the disease with symptomatic improvement for three months. He responded to adriamycin therapy in spite of resistance to dactinomycin and vincristine treatment.

Case 2: F.W. A two-year-old male was found to have a large abdominal mass in February 1970 and a diagnosis neuroblastoma was made on a laparotomy in March 1970. At that time he had superior mediastinal widening in addition to the abdominal mass. Bone marrow aspiration revealed occasional extrinsic cells forming clumps. He was treated with radiotherapy, cytoxan, vincristine, and procarbazine with minimal response. In July 1970 he developed a large scrotal mass and evidence of progressive disease. A course of adriamycin in July 1970 produced marked regression of scrotal, abdominal, and mediastinal mass. After recovery from leukopenia, he received the drug on a maintenance schedule. Remission lasted for two months before recurrent disease became evident.

Comment: This two-year-old boy with wide-spread metastatic neuroblastoma showed a remarkable response to the first course of adriamycin in spite of resistance to cytoxan, vincristine, and procarbazine.

Case 3: G. D. This 17-year-old white male was diagnosed as having an osteogenic sarcoma of the right femur in June 1970. He was initially treated with right hip

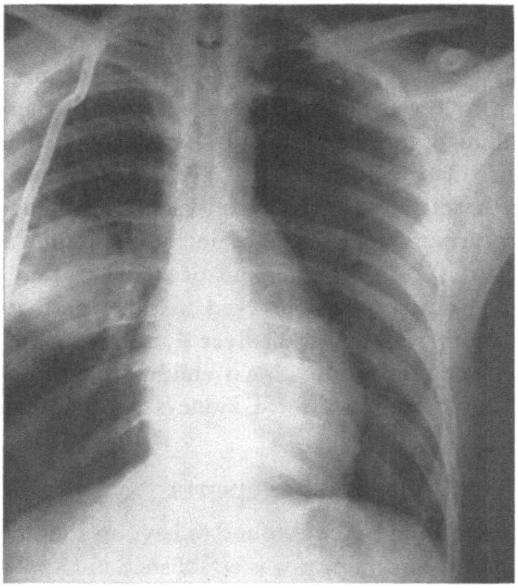

Fig. 1a. Chest X-ray of case 3, showing pulmonary metastasis from an osteogenic sarcoma before adriamycin therapy

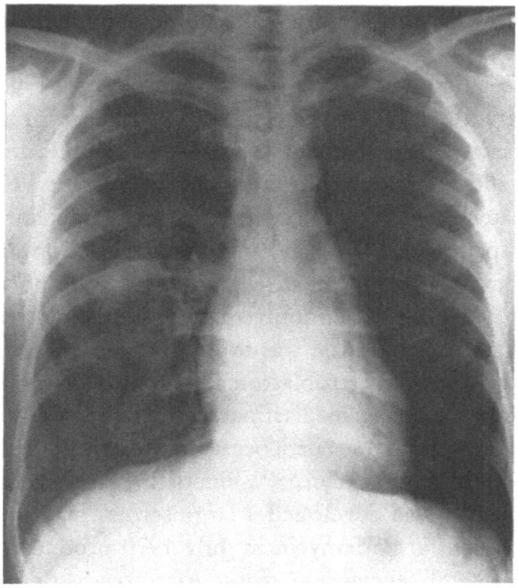

Fig. 1b. Chest X-ray of case 3, showing partial regression of pulmonary mass after adriamycin therapy

disarticulation and remained free of disease until September 1970, when a right pulmonary metastasis was noted. In October 1970, adriamycin 35 mg/m^2/day for three consecutive days was initiated. The pulmonary mass regressed after the first course of adriamycin therapy and the lesion has remained static for five months while receiving maintenance adriamycin therapy (Fig. 1a, 1b).

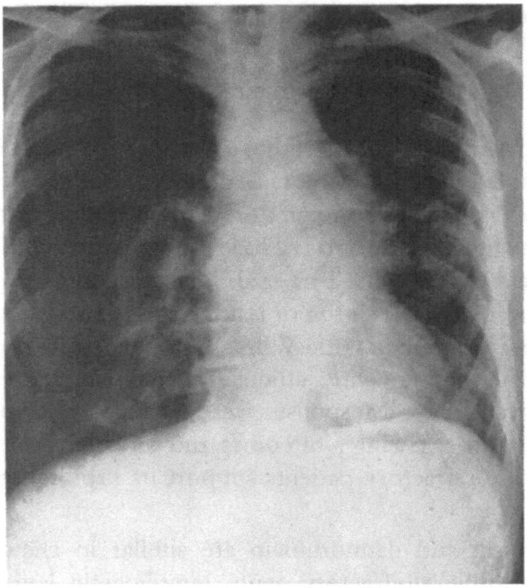

Fig. 2a. Chest X-ray of case 4, showing a bronchogenic carcinoma (anaplastic) of left lung before adriamycin therapy

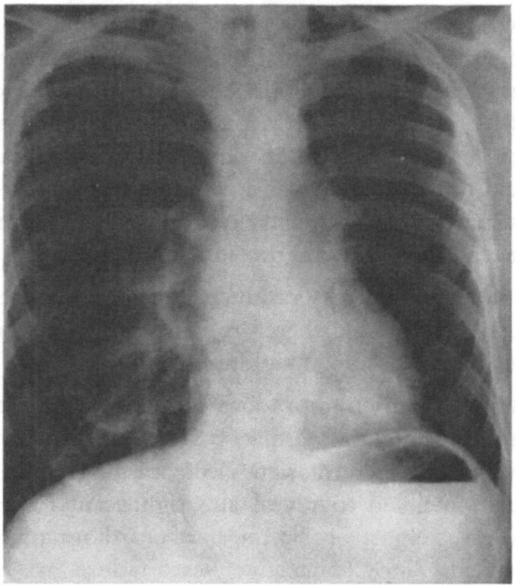

Fig. 2b. Chest X-ray of case 4, showing partial regression of left hilar and pulmonary mass after adriamycin therapy

Case 4: K. R. A 64-year-old white male presented with three months history of cough and hemoptysis. In November 1970, chest X-rays revealed left hilar lymphadenopathy and a mass in the legt lung. Cytology of sputum confirmed the diagnosis of an anasplastic bronchogenic carcinoma. Because the tumor was unresectable, he was treated with adriamycin 35 mg/m² daily for three consecutive days. Regression of hilar and pulmonary mass was noted after the first course of therapy. He has been followed for four months with no evicence of regrowth (Fig. 2a, 2b).

Discussion

In this study, adriamycin appears to have a wide spectrum of antitumor activity against childhood and adult solid tumors. Adriamycin induced remission in Wilms' tumor, neuroblastomas, Ewing's sarcoma, reticulum cell sarcomas, osteogenic sarcomas, and bronchogenic carcinomas. These findings are consistent with reports from Bonadonna et al., [3,4] and Tan et al., [5]. Response of osteogenic sarcoma to adriamycin is of interest since this tumor is nonresponsive to previously tested drugs. One child with progressive metastatic Wilms' tumor, despite therapy with dactinomycin and vincristine, subsequently attained a complete regression of metastatic disease with adriamycin. Partial responses were seen also in cytoxan and vincristine refractory neuroblastomas, Ewing's sarcomas and reticulum cell sarcomas. Activity fo adriamycin in these refractory patients support its exploration in patients earlier in their therapeutic course.

Althoug adriamycin and daunorubicin are similar in chemical structure, two patients with daunorubicin-refractory acute lymphocytic leukemia responded to adriamycin suggesting independent pharmacological characteristics [2].

Major toxicity of adriamycin included myelosuppression, mucosal ulceration and alopecia. Myelosuppression was limiting toxicity, usually occurring between the second and third week and lasting for about seven to ten days before recovery. Severe myelosuppression, often associated with fever and stomatitis, requires intensive supportive care. The nadir of both leukocyte and platelet counts occurred earlier in patients with leukemia than in patients with solid tumor. This may be explained by the better bone marrow reserve in the latter. Alopecia, usually appearing between the third and fourth weeks, was a common but not dose limiting toxicity. As observed by Bonadonna et al. adriamycin produced a more pronounced alopecia than did daunorubicin. Hair usually regrew in 2—4 months.

Adriamycin is a vesicant and may produce severe chemical cellulitis if the drug is given paravenously. It should, therefore, preferably, be injected by separate syringe through a scalp vein needle in children.

Daunorubicin has been known to produce cardiac toxicity therefore, close attention was directed to the potential cardiac toxicity of adriamycin. Four patients treated with adriamycin developed transient electrocardiographic changes. However, post-mortem examination failed to reveal any significant findings. The temporal sequence is, nevertheless, suggestive that the electrocardiographic changes should be attributed to adriamycin. Since congestive heart failure has been reported in a child treated with adriamycin it is suggested that the total dose of adriamycin should not exceed 600 mg/m² body surface.

Summary

Adriamycin is a hydroxymethyl [1] analog of daunorubicin with antitumor effect in man. We have studied 86 patients: 44 with acute leukemia (27 children and 17 adults), and 42 solid tumors (20 children and 22 adults). Objective response was seen in 1 Wilm's tumor, 5 of 8 neuroblastomas, 2 of 3 Ewing's sarcomas, 4 of 5 reticulum cell sarcomas, 5 of 13 osteogenic sarcomas, 3 of 6 bronchogenic carcinomas, and 1 of 1 lymphosarcoma. Complete remission (M-1 marrow) was seen in 1 of 14 acute myelocytic leukemias. Of 30 acute lymphocytic leukemias, 5 had complete and 1 had partial remission. Responses lasted from 0.25 months to 10+ months with a median of 3 months. Except for patients with osteogenic sarcomas and bronchogenic carcinomas, all received prior chemotherapy. Major toxicity of adriamycin included granulocytopenia, lymphocytopenia, thrombocytopenia, anemia, marrow hypoplasia, stomatitis, vomiting and alopecia. Activity of adiamycin in these patients with advanced disease supports its exploration in patients earlier in their therapeutic course.

Acknowledgments

Sincere appreciation is extended to members of the clinical staff at the Departments of Medicine A and Pediatrics of Roswell Park Memorial Institute who helped in these trials.

References

1. Di Marco, A., Gaetani, M., Scarpinato, B.: Cancer Chemother. Rep. **53**, 33—37 (1969).
2. Sandberg, J. S., Howsden, L., Di Marco, A., Goldin, A.: Cancer Chemother. Rep. **54**, 1—7 (1970).
3. Bonadonna, G., Monfardini, S., De Lena, M., Fosati-Bellani, F.: Brit. med. J. 1969 III, 503—506.
4. Bonadonna, G., Monfardini, S., De Lena, M., Fosati-Bellani, F., Beretta, G.: Cancer Res. **30**, 2572—2582 (1970).
5. Tan, C., Wollner, N., King, O., Ilano, D.: Proc. Amer. Ass. Cancer Res. **10**, 79 (1970).

Adriamycin in Children with Acute Leukemia and Other Neoplastic Diseases[1]

CHARLOTTE TAN, ERLINDA ETCUBANAS, NORMA WOLLNER, GERALD ROSEN, M. LOIS MURPHY, and IRWIN H. KRAKOFF

Division of Chemotherapy Research, Sloan-Kettering Institute for Cancer Research, and Department of Pediatrics, Memorial Hospital for Cancer and Allied Diseases, New York, USA

With 3 Figures

At Memorial Hospital, New York, 172 children with leukemia and various types of metastatic cancer have been treated with adriamycin. These included the patients treated, and summarized in two abstracts [1, 2]. 51 patients with acute leukemia, and 67 with solid tumors received adriamycin alone. The remaining 54 children were given adriamycin either sequentially or in combination with other chemotherapeutic agents, as a multiple drug protocol.

This report includes the results of adriamycin alone in the 118 children, and in a multiple drug combination, in the 12 children with Hodgkin's disease.

Table 1 shows the six intravenous dose schedules used. Some patients received more than one dose schedule. Initially, adriamycin was given at 0.5 mg/kg by single

Table 1. Adriamycin in 172 children intravenous dose schedules

Dose (mg/kg) Single	Total/course	Days/course (Schedule)		No. of patients	Average No. of courses patient (range)
(1)[a] 0.5	2–3	4– 6	(C)	102	2 (1–15)
(2) 0.5	2–9	8–74	(I)	73	2 (1–5)
(3) 0.25–0.5	1–2	1	(q 6 h × 4)	16	3 (1–7)
(4)[a] 1.0	2–3	2– 3	(C)	18	2 (1–9)
(5) 1.0	2–4	5–10	(I)	9	1 (1–2)
(6) 2.0	2	1		4	2 (1–3)

[a] Recommended dose schedules (1) or (4).

Schedule:	Maximum total dose mg/kg (months)		
C = consecutive days	Leukemia	19	(6)
I = intermittent days	Solid tumor	40	(14)

1 Supported, in part, by Grants CA-08748 and CA-08526 of the National Cancer Institute.

injection, to a total of 2—3 mg/kg per course on 4—6 consecutive days. Courses may be repeated in 3—4 weeks. One hundred two patients had these schedules, with an average of two, but ranged from one to 15 courses per patient. 73 patients had the same single dose, but on intermittent days. Sixteen patients were given 0.25 to 0.5 mg/kg every 6 h for four doses. Single doses of 1 mg/kg, either consecutively or or intermittently were given to 27 patients. Four patients had 2 mg/kg in a single dose. No detectable differences in toxicity were noted; however, the recommended schedules were one or four, with the drug given on consecutive days. The maximum total dose in a patient with leukemia was 19 mg/kg, given in 6 months, and solid tumors, 40 mg/kg given in 14 months.

Toxic Effects of Adriamycin

Hematopoietic Depression. At an average total dose of 3 mg/kg/course, 64% of the children with solid tumors developed leukopenia and thrombocytopenia, with occasional bleeding and anemia. The hematopoietic depression usually reached a nadir in two weeks, and recovered by the third week.

Alopecia. 87% of the children developed alopecia, which began five to 30 days after the start of treatment. Regrowth usually occurred within two months, after the discontinuation of the drug.

Mouth Lesions. 71% of the children developed oral ulcers. These usually occurred between ten to 18 days after the start of treatment, and lasted for three to seven days.

Fever. This usually coincided with the maximal hematopoietic depression, and occurred in 52% of the patients. In general, fever appeared to be due to infection, secondary to leukopenia, and not a drug-related phenomenon.

Gastrointestinal Disturbances. Most children tolerated a single dose of 0.5 mg/kg. A higher single dose of 1 or 2 mg/kg may cause more nausea, vomiting, or abdominal pain.

Local Reaction. Extravasation of adriamycin may cause severe local cellulitis, followed by necrosis and fibrosis of the subcutaneous tissue.

Others. All patients had electrocardiographic studies before and after adriamycin. A collective review of the electrocardiograms in 26 children who received a total cumulative dose of 12—40 mg/kg of adriamycin in 3—14 months, showed that in eight patients there were varying degrees of changes, which consisted of prolonged ST segments, and flattening of T-waves, and in some, a decreased QRS voltage. These findings have returned to normal within one to two months, after the discontinuance of adriamycin. In one child with Wilms' tumor, after two courses of adriamycin, her pulmonary metastases disappeared. Electrocardiographic changes occurred during the fourth course, and progressively became worse. She developed heart failure after the ninth course. Her total dose of adriamycin was 24.5 mg/kg. She died; unfortunately, no autopsy was obtained.

Therapeutic Responses

The response of acute leukemia to treatment was evaluated according to the "Criteria for Evaluating Chemotherapy in Acute Leukemia" by the Children's Cancer Study Group A [3]. Four criteria of response were measured: bone marrow (M), hemogram (H), physical examination (P), and symptoms of performance status (S). A complete hematopoietic and clinical remission (CR) was classified M1–H1—P1—S1 and a good partial remission (GPR) was M1 or M2, H1 or H2, P1 or P2, and S1 or S2.

Adriamycin Alone

In acute leukemia (Table 2), 16 of the 38 patients with lymphoblastic leukemia adequately treated, had complete and good partial remissions, lasting for 2—5 months (Case 1). Three of the five non-lymphoblastic leukemias had shorter remissions (Case 2).

Table 2. Adriamycin remissions in 51 children acute leukemias

	Number of patients		Remission (mos.)	
	Total	Adequate	Complete	Good partial
Lymphoblastic	42	38	5 (2–5)	11 ($^1/_4$–$2^1/_2$)
Myelo, monoblastic	7	5	2 ($^3/_4$, 1)	1 ($^3/_4$)
Miscellaneous	2	1	—	—

Case 1: J.M. A $3^1/_2$ year old boy with acute lymphoblastic leukemia, diagnosed in July 1968, was in his third marrow relapse in April 1970, when adriamycin was begun. He achieved a complete remission, which was maintained with bi-weekly doses, for a total of five months; and while in remission, he was placed on other maintenance drugs.

Case 2: C.M. A $3^1/_2$ year old girl with acute monoblastic leukemia, diagnosed in November 1970, had an initial leukocyte count of 260,000/cu mm, with 90% monoblasts in the bone marrow. She was given adriamycin 0.5 mg/kg every six hours for four doses, a total of 2 mg/kg in 24 h, followed by a precipitous drop in her leukocyte count, and decrease in the percentage of blasts in the bone marrow. A similar dose was repeated every two weeks. She achieved a complete remission until January 1971, when she had 15% blasts in the bone marrow, and she was then placed on arabinosylcytosine and thioguanine. There was no further remission.

Tumor responses were classified according to the Karnofsky Categories of Response [4].

Category O—O: Disease progresses.

O—A: Subjective benefit without favorable objective changes.

O—B: Favorable objective changes without subjective benefit.

O—C: Subjective benefit and favorable objective changes in measurable criteria, but of less than one month duration.

Category I—A: Distinct subjective benefit with favorable objective changes in all measurable criteria for one month or more.

I—B: Objective regression of all palpable or measureable neoplastic disease (all lesions be reduced at least 50% in bulk), for one month or more in a relatively asymptomatic patient.

I—C: Complete relief of symptoms, if any, and regression of all manifestations of disease for one year or more.

Category II: Interruption or slowing in the progression of the disease without definite evidence of subjective or objective improvement.

In the patients with solid tumors and lymphomas (Table 3), all had metastatic or recurrent disease, and the majority had other drugs previously, such as dactinomycin, vincristine, and some had daunomycin; 44 of 57 (78%) of the patient had tumor regressions with adriamycin. Of the adequately treated, 5 of the 14 with embryonal rhabdomyosarcoma had a I—A response lasting for one to eight months (Case 3).

Table 3. Adriamycin responses in 67 children solid tumors and lymphomas

Diagnosis	Number of patients		Response categories (mos.)						
	Total	Adequate	I–C	I–A	I–B	O–A	O–B	O–C	II
Embryonal rhabdomyosarcoma	16	14		5(1–8)				6	1
Neuroblastoma	12	10		1(2)		1		2	
Hodgkin's disease	10	9		4($1^1/_2$–3)			1	4	
Ewing's sarcoma	7	6	1(13+)	1($3^1/_2$)	1(1)		2	1	
Wilms' tumor	5	5		2(2, $3^1/_2$)	1(10)				1
Ovarian tumors	4	3	1(13+)	2($4^1/_2$, 5)					
Lymphosarcoma/ reticulum cell sarcoma	4	3	1(6)				1	1	
Hepatoma	3	3						1	
Embryonal carcinoma	2	2		1(4+)					
Malignant teratoma (sacral)	1	1			1(1)				
Miscellaneous	3	1							

Only one of the ten neuroblastoma patients had a I—A response. The majority of this group had had courses of daunomycin previously, to which there is probably cross resistance, and no response could be anticipated. Four of the nine Hodgkin's disease patients had I—A responses lasting for $1^1/_2$ to three months. Of the six Ewing's sarcoma, one had a I—C response for 13+ months (Case 4); two had I—A and I—B responses for $3^1/_2$ and one month respectively. Three of the five Wilms' tumor had Category I response lasting for two to ten months (Case 5). Of the three patients with ovarian tumors, one had a I—C response for 13+ months, and two had I—A responses for $4^1/_2$ and five months (Case 6) respectively. One of each with lymphosarcoma and embryonal carcinoma, had I—A responses for six and 4+ months respectively. The Categories O—A, B, and C responses are as listed in the tables. One of each with embryonal rhabdomyosarcoma and Wilms' tumor had Category II responses, indicating no progression or regression of disease during adriamycin treatment.

Case 3: R.M. A $3^1/_2$ year old girl had pain and swelling in the right parotid region in May 1970. In July the mass measured 8 × 8 cm, associated with right facial palsy. Biopsy showed embryonal rhabdomyosarcoma. X-rays showed destruction of the right mandible, and bilateral pulmonary nodules. A course of five doses of adriamycin was given in July 1970. A week later the pain disappeared and the tumor mass decreased to 5 cm in size, with complete resolution of the facial palsy. By August, chest x-ray showed resolution of the pulmonary nodules, and remained clear while she was given monthly courses of adriamycin. In September the tumor mass over the mandible was no longer palpable, followed by decrease in the bony destruction in the mandible. However, in November 1970, 5 months after starting adriamycin, the mass recurred. It was excised and showed recurrent embryonal rhabdomyosarcoma. She was given 6000 rads to the right parotid region in a 6.5 cm field. In January 1971 x-ray of the mandible showed complete healing of the former destruction. In April a small soft tissue mass recurred again. Wedge resection was done, which was embryonal rhabdomyosarcoma. She is now on a multiple drug protocol.

Case 4: M.M. A 10 year old boy with Ewing's sarcoma of the right iliac bone, diagnosed in December 1969, was given irradiation to the primary site. 6 months later, he developed bilateral pulmonary metastases, with right pleural effusion. Adriamycin was begun on 6/5/70, and he completed a course by 6/11/70. At that time there was no favorable change in his x-rays. 6 days after the completion of the first course, there was complete clearing of the effusion and pulmonary metastases (Fig. 1). He is now more than 14 months after the onset of pulmonary metastases, and is being treated with monthly courses of adriamycin, with no recurrence of disease in the primary site, or pulmonary metastases.

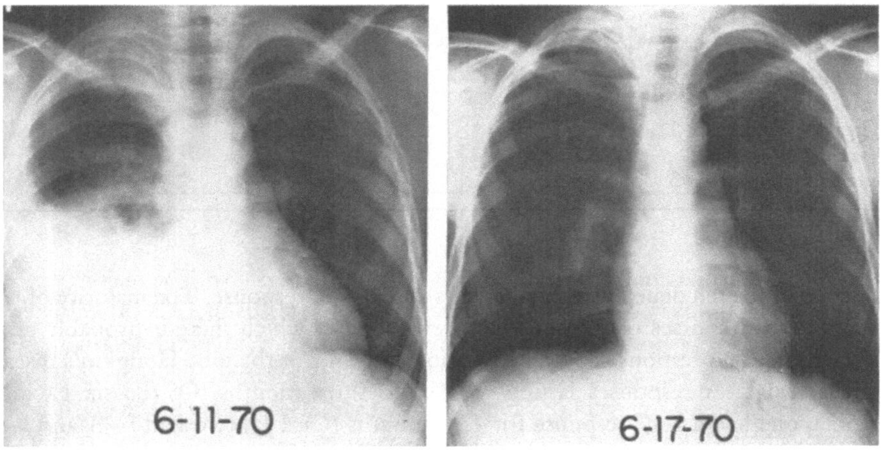

Fig. 1. Chest X-ray of M.M. (Case 4) done on 6/11/70, and on 6/17/70, after a course of adriamycin

Case 5: S.Q. A 4 year old girl with Wilm's tumor had a left nephrectomy, postoperative abdominal irradiation, and received dactinomycin, vincristine, as well as irradiation to the lungs for bilateral pulmonary metastases. In August 1969, an attempt to excise the left recurrent lung metastases was unsuccessful. After 2 courses of adriamycin, there was more than 50% tumor regression, which lasted for $3^1/_2$ months (Fig. 2). There was no further improvement on additional courses.

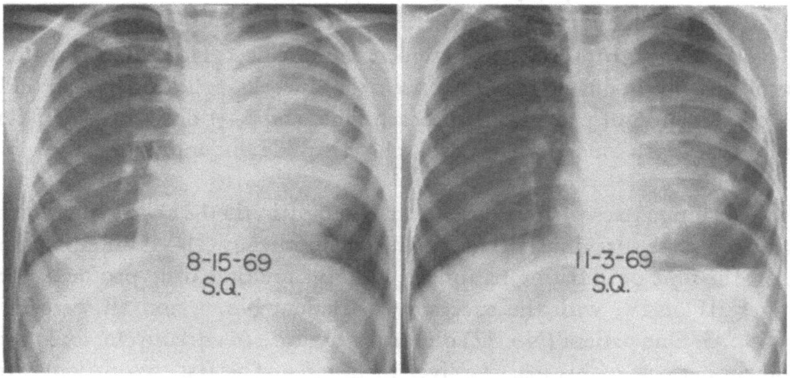

Fig. 2. Chest X-ray of S.Q. (Case 5) done on 8/15/69 before therapy, and on 11/3/69 after adriamycin therapy

Case 6: J.A. A 9 year old girl had a left ovarian tumor removed in March 1970. Diagnosis was embryonal carcinoma of the ovary. 6 months later she had a sudden onset of chest pain. X-ray showed a left hilar mass. Adriamycin was begun in September 1970. The mass in the left lung completely disappeared, as shown in the x-ray or January 1971 (Fig. 3); however, in February she developed headache and vomiting, followed by convulsions. Cerebral angiogram and brain scan showed left cerebellar and medial parieto-occipital metastatic lesions which were irradiated.

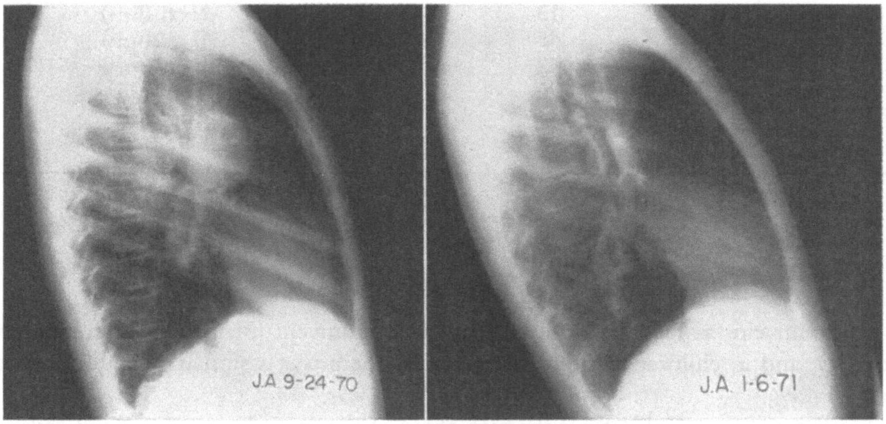

Fig. 3. Chest X-ray of J.A. (Case 6) done on 9/24/70 before therapy, and on 1/6/71 after adriamycin therapy

She was continued on monthly courses of adriamycin with no recurrence of the lesion. Unfortunately, she died at home on 8/9/71, of cause unrelated to her cancer, according to the local physician.

Adriamycin in Combination

A multiple drug protocol was designed for advanced Hodgkin's disease, with a course of adriamycin during the first phase of 3—4 weeks, followed by a second phase, with a combination of prednisone, procarbazine, and 3—4 doses of vincristine, and a third phase with one dose of cyclophosphamide in approximately a 3 month period, then re-cycle.

Table 4 summarizes the response to this therapy in the 12 children with advanced Hodgkin's disease, previously treated with chemotherapy, 1—45 months after diagnosis. At the time of initiating treatment with this multiple drug protocol, they were all Stages III or IV, with the exception of patient No. 6, and all were B, except patient No. 3. One patient (No. 12) died after a course of adriamycin, due to toxicity, most likely because of his massive liver metastases. The last two patients were too early for evaluation. All 9 patients have had tumor regressions: four patients for more than a year, and 5 for 3 to 7+ months.

Table 4. Adriamycin in multiple drug protocol for advanced Hodgkin's disease in children

Patient No.	Previous therapy R.T.	Previous therapy Chemo.	Diagnosis to start of M.D.P. (mos.)	Stage at Dx.	Stage at M.P.D.	Response category (mos.)
1	+	—	18	II	IV	I—C (15+)
2	—	—	1	IV	IV	I—C (14+)
3	+	+	37	III[a]	III	I—C (12)
4	+	+	7	III	IV	I—C (12+)
5	+	—	19	II	III	I—A (2¾)
6	—	—	2	II	II	II (7+)
7	+	+	13	III	IV	I—A (6+)
8	+	+	45	III	IV	II (6+)
9	+	+	42	IV	IV	I—A (5+)
10	+	+	26	II	III	Too early
11	+	+	44	II	IV	Too early
12	+	+	8	III	IV	I

[a] All "B" except patient No. 3.

Discussion

Adriamycin, at half the dose of that of daunomycin [5], produced comparable toxicity, and in children with acute leukemia, there was a similar overall remission rate of 43%.

In the non-lymphoblastic leukemias, three of the five adequately treated children (including the two previously untreated) obtained complete or good partial remissions. Althongh this number is too small to make a comparison with other chemotherapeutic agents, the observed effectiveness of adriamycin alone leads us to believe that adriamycin has a role in the treatment of non-lymphoblastic leukemia, either as a single agent, or in combination with other agents.

In general, as with daunomycin, adriamycin produced a rapid remission, but the duration of remission was usually short. At Memorial Hospital, adriamycin is now being given in combination with prednisone and vincristine, in some patients with acute lymphoblastic leukemia for remission induction, and in non-lymphoblastic leukemia, as consolidation after initial induction with arabinosylcytosine and thioguanine.

In children with solid tumors and lymphomas, the overall tumor response rate was 78% which was higher than that observed with daunomycin. Adriamycin produced tumor regression in a wide variety of tumors, especially Hodgkin's disease, lymphosarcoma, reticulum cell sarcoma, other soft tissue sarcomas, Ewing's sarcoma, and the less commonly seen tumors, such as ovarian tumor, embryonal carcinoma, and malignant teratoma.

5 of the 10 children with neuroblastoma treated with adriamycin had had course of daunomycin previously, to which there is probably cross resistance. This may explain why only three patients had I—A and O—C responses. However, clinically adriamycin was not cross resistant with other chemotherapeutic agents, such as dactinomycin, vincristine, or cyclophosphamide. BONADONNA [6] and WANG [7] have documented the therapeutic effects of adriamycin in neuroblastoma; a total of ten of 14 treated responded. This response rate is similar to our experience with daunomycin.

The duration of tumor regression in the majority of solid tumor patients was also short. It lasted for less than one month in 46%, and from 1—6 months in 41%. The remaining 12% of the patients had longer periods of tumor response for 7—14+ months. The reasonable approach in the treatment of malignant neoplasms then, is to use adriamycin in combination with other effective drugs of different mechanisms of action. This is now being studied at Memorial Hospital.

DEVITA et al. [8] and LOWENBAUN et al. [9], reported combination chemotherapy with nitrogen mustard, vincristine, procarbazine, and prednisone in patients with Hodgkin's disease, resulted in a complete response rate of 81% in the previously untreated, and a complete response rate of 70% in the previously treated patients. The long durations of response were encouraging.

In our initial trials with adriamycin alone, all nine children with advanced Hodgkin's disease had tumor regressions; however, the durations were short. It seemed reasonable then, to maintain the therapeutic effects obtain through adriamycin, with other effective agents, such as prednisone, procarbazine, vincristine, and cyclophosphamide. It is too early to determine the effectiveness of this multiple drug protocol, using adriamycin in the initial phase.

In children, the results of treatment with adriamycin are particularly encouraging, since adriamycin has demonstrated a wide spectrum of activity against most forms of solid tumors of childhood. In addition, once tumor response to adriamycin has occurred, it appears that treatment can be continued for longer periods of time than with daunomycin, because the toxic effects were more predictable, and so far the cardiac toxicity does not occur as readily at a therapeutic dosage of adriamycin. However, the fact that the majority of tumors which responded to adriamycin eventually recurred, leads us to believe that the chemotherapy with a variety of effective drugs used in rotation or in combination, is preferable to single drug therapy. The finding of a drug such as adriamycin, with a high therapeutic index in a wide spectrum of childhood tumors, makes it a welcome addition to the chemotherapeutic agents available for the treatment of neoplastic diseases.

Summary

Adriamycin produced a 43% overall remission rate in previously treated patients with acute leukemia. This is similar to the study we previously reported with daunomycin.

In solid tumors and lymphomas, 78% had tumor regression. Adriamycin has a rather wide range of activity, not only in the common types of malignant neoplasms of childhood, such as embryonal rhabdomyosarcoma, neuroblastoma, Wilms' tumor, Ewing's sarcoma, and lymphoma, but also causes tumor regression in the less common types, such as malignant teratoma, and embryonal carcinoma.

With the monthly course of adriamycin, the toxic effects have been minimal, although there have been varying degrees of transient electrocardiographic changes during, or immediately after treatment. Adriamycin may have been the cause of death in one child, who died of heart failure after nine months of treatment, but who had also received other chemotherapeutic agents, and pulmonary irradiation for recurrent metastatic Wilms' tumor.

The results in Hodgkin's disease, so far, have been encouraging, with the addition of adriamycin to other effective drugs, but it will require a comparative study of precise experimental design, and large numbers of patients, such as the National Cooperative Group Studies, to make a conclusion.

References

1. Tan, C., Wollner, N., King, O., Ilano, D.: Proc. Amer. Ass. Cancer Res. **11**, 79 (1970).
2. Wollner, N., Tan, C., Ghavimi, F., Tefft, M., Murphy, M. L.: Proc. Amer. Ass. Cancer Res. **12**, 75 (1971).
3. Hewlett, J., Battle, J., Bishop, R., Fowler, W., Schwartz, S., Hagen, P., Louis, J.: Cancer Chemother. Rep. **42**, 25 (1964).
4. Karnofsky, D. A.: Clin. Pharmacol. Ther. **2**, 709 (1961).
5. Tan, C., Tasaka, H., Yu, K., Murphy, M. L., Karnofsky, D.: Cancer **20**, 333 (1967).
6. Bonadonna, G., Monfardini, S., DeLena, M., Fossati-Bellani, F., Beretta, G.: Cancer Res. **30**, 2572 (1970).
7. Wang, J., Cortes, E., Sinks, L., Holland, J.: Cancer **28**, 837 (1971).
8. DeVita, V., Serpick, A., Carbone, P.: Proc. Amer. Ass. Cancer Res. **10**, 19 (1969).
9. Lowenbraun, S., DeVita, V., Serpick, A.: Blood **36**, 704 (1970).

Adriamycin in the Treatment of Acute Leukemia[1]

J. M. A. Whitehouse, D. Crowther, and J. S. Malpas

Department of Medical Oncology, St. Bartholomew's Hospital, West Smithfield, London E.C.L./Great Britain

The report by Bonadonna and his colleagues in 1969 of the effectiveness of the anti-tumor antiobiotic adriamycin in a variety of malignant conditions including those affecting children suggested that a trial in acute leukemia should be undertaken. A total of 29 patients have so far been treated who have been in relapse with either acute lymphoblastic leukemia, acute myelogenous leukemia or lymphoblastic lymphosarcoma. 10 patients had acute lymphoblastic leukemia, 13 acute myelogenous leukemia and 6 had lymphoblastic lymphosarcoma (Table 1).

Because of the efficacy of vincristine and prednisolone in combination in acute lymphoblastic leukemia, Hardisty et al. [2] (1969), and daunorubicin and cytosine arabinoside, Crowther et al. [3] (1970), in acute myelogenous leukemia these regimes had always been employed first so that patients receiving adriamycin had either failed to go into remission or were in a second or subsequent relapse. In addition the acknowledged effectiveness of combination therapy made us reluctant to use adriamycin alone so that in many cases of acute lymphoblastic leukemia prednisolone was used concurrently and in acute myelogenous leukemia adriamycin was given with cytosine arabinoside, thioguanine or 6 mercaptopurine (Table 1).

Adriamycin was given in courses of 2—3 days using a dose of 0.4 mg/kg/d. The dose was given by one intravenous injection daily. Intervals of 7—10 days were left between courses. The total dose of adriamycin administered varied between 20 and 510 mg and it would appear that between 100 and 150 mg has to be administered to induce remission in these cases.

Because of previous experience with the anti-tumour antibiotic daunorubicin, Malpas and Scott [4] (1968), Malpas and Scott [5] (1969), frequent surveillance of the peripheral blood count and electrocardiograph was carried out while the patient was being treated. A complete remission was established when the patient was without symptoms or signs, had a normal peripheral blood with no circulating blast-cells and when a bone marrow examination showed less than 5% of blast cells present.

Results

Details of the patients are given in Table 1 and summarised in Table 2.

1 Adriamycin was kindly supplied by Pharmitalia U.K.

Table 1. Patients with acute leukemia treated with adriamycin

Case no.	Age and sex	Diagnosis	Previous therapy	Concurrent therapy	Total dose mg	Result	Comments
1. J.V.	14 M	ALL	P.VCR.MTX	P. AraC.	140	PR	
2. K.B.	18 F	ALL	P.VCR.MTX	AraC.	480	CR	2 successive CR's
3. R.S.	5 M	ALL	P.6MP.MTX VCR		30	DIED	
4. V.W.	47 F	ALL	P.VCR.AraC DR. MTX		150	PR	CR then failed DR therapy
5. A.M.	17 M	ALL	P.VCR.AraC DR.	P	300	PR	
6. D.McD.	38 M	ALL	P.6MP.VCR. CY. DR.	P. 6MP	90	F	Hypoplastic before A
7. E.A.	58 F	ALL	P.VCR.MTX 6MP.As	P	130	F	
8. D.H.	12 M	ALL	VCR.P.MTX 6MP.AraC DR. As	T with 2nd CR	510	CR	2 successive CR's on A after 2 CR's with DR
9. K.F.	12 M	ALL	6MP. P.VCR. CY. MTX. AraC. DR.	P	140	F	Hypoplastic
10. S.F.	20 F	ALL	P.VCR. failed	P	150	CR	
11. R.McQ.	F	L/LSA	P.VCR.	P	180	F	Spleen reduced in size
12. A.C.	15 F	L/LSA	P.VCR.DR. AraC.MTX	P	300	CR	Stomatitis
13. P.A.	22 M	L/LSA	P.VCR.DR.	T	150	F	Hypoplastic
14. D.W.	26 M	L/LSA	HN2 N.P.VCR	T	20	F	Died with staph septicemia but not pancytopaenia due to A
15. E.W.	25 F	L/LSA	P.VCR.	AraC	60	PR	
16. B.R.	41 F	L/LSA	DXR	AraC	120	PR	
17. W.E.	61 M	AML	DR. AraC	AraC	50	PR	
18. B.H.	34 M	AML	6MP. P. As. DR.AraC. T.	AraC	100	F	
19. J.L.	13 F	AML	DR. AraC	6MP	60	F	
20. I.L.	38 F	AML	DR. AraC As.	6MP	150	F	
21. W.Q.	45 F	AML	P.VCR.6MP DR. AraC	AraC	120	F	Reduction in the blasts in the bone marrow
22. G.G.	15 M	AML	DR. P. VCR. AraC	—	60	F	Fall in blast count
23. T.F.	21 M	AML	VCR. P. AraC DR. MTX	T	225	F	
24. M.F.	58 M	AML	DR. AraC	T	60	F	Fall in blast count

Table 1 (continued)

Case no.	Age and sex	Diagnosis	Previous therapy	Concur-rent therapy	Total dose mg	Result	Comment
25. B.R.	36 M	AML	DR. AraC 6MP MTX CY. P	T	60	F	
26. C.G.	54 M	AML	P. CY. DR. AraC MTX	T	160	F	
27. S.A.	71 F	AML	DR. AraC	T	120	F	Fall in blast count
28. A.W.	73 F	AML	DR. AraC	T	50	F	
29. P.O's	21 F	AML	As. DR. AraC	6MP	60	F	

A = Adriamycin. As = Asparaginase. AraC = Cytosine arabinoside. Cy = Cyclophosphamide. DR = Daunorubicin. 6MP = Mercaptopurine. MTX = Methotrexate.

P = Prednisolone. T = Thioguanine. N = Natulan. HN2 = Nitrogen mustard. VCR = Vincristine. DXR = Radiotherapy. CR = Complete remission. PR = Partial remission. F = Failure.

ALL = Acute lymphoblastic leukaemia. AML = Acute myeloblastic leukaemia. L/LSA = Lymphoblastic lymphosarcoma.

Table 2. Adriamycin in acute leukemia

Diagnosis	Number of patients	Results
Acute lymphoblastic leukemia	10	3 CR 3 PR 4 failed
Acute myeloblastic leukemia	13	1 PR
Lymphoblastic lymphosarcoma	6	1 CR 1 PR

Remission Rates

Of the 10 patients with acute lymphoblastic leukemia 3 achieved complete remission. It is of interest that in 2 of these patients a further complete remission was obtained (Case 2 and Case 8). Furthermore, one of these patients had already had a complete remission induced by daunorubicin. In another patient 3 courses of vincristine and prednisolone had failed to induce remission so that adriamycin was substituted for vincristine (Case 10). A severe marrow hypoplasia was induced by 150 mg of adriamycin but nevertheless a good complete remission was obtained.

In one patient with acute lymphoblastic leukemia (Case 2), a complete remission was obtained rapidly with 120 mg of adriamycin combined with a course of cytosine arabinoside, a combination which appears to be suitable for further investigation.

The results of treatment were disappointing in the 13 patients with acute myelogenous leukemia. One patient only achieved a partial remission (Case 17).

In the patients with lymphoblastic lymphosarcoma one achieved a complete remission (Case 11), and one achieved a good partial remission (Case 16).

Side Effects and Major Toxicity

Adriamycin has a rapid and marked effect on the peripheral blood count. Although quite severe pancytopenia and marrow hypoplasia was seen in the patients treated, no deaths attributable to haemorrhage or sepsis occurred, none of the patients was treated in pathogen free areas. Serial electrocardiograms showed no evidence of toxicity even in the elderly patients treated for myelogenous leukemia. In the child who did show ECG changes these were already present after the course of daunorubicin. No further changes were produced by two courses of adriamycin. Alopecia and stomatitis occurred but the stomatitis was never troublesome and in the dose of adriamycin used never gave rise to necrotic lesions of the mouth reported by MATHÉ et al. [6] (1970).

Discussion

Adriamycin can induce remission in acute lymphoblastic leukemia and lymphoblastic lymphosarcoma. In acute lymphoblastic leukemia it forms a very useful second-line drug which may be substituted for vincristine in the event of the failure of remission induction by conventional methods or it may be used when relapse has occurred. In one of our cases cross resistance between daunorubicin and adriamycin was not apparent for after a further attempt to achieve remission with daunorubicin failed, a rapid and successful complete remission was achieved with adriamycin. Thit one case will require confirmation.

The disappointing results in acute myelogenous leukemia would suggest that either the present regime was not sufficiently intensive to achieve remission or the fact that all the patients have previously received daunorubicin resulting in the recurring myeloblasts being insensitive to the second drug. Certainly MATHÉ et al. [4] were able to obtain 6 complete remissions in 18 patients with myelogenous leukemia but noted that 4 patients previously treated with daunorubicin failed to benefit from adriamycin.

A comparative trial of adriamycin and daunorubicin in acute myelogenous leukemia will be necessary however to finally decide this question.

Acknowledgments

We gratefully acknowledge the help of Dr. C. J. T. BATEMAN and Dr. M. BEARD with the haematological management of these cases and the advice of Sir RONALD BODLEY SCOTT and Dr. G. HAMILTON FAIRLEY.

References

1. BONADONNA, G., MONFARDINI, S., DELINA, M., FOSSATI-BELLANI, F.: Brit. med. J. **1969 III**, 503.
2. HARDISTY, R. M., MCELWAIN, T. J., DARBY, C. W.: Brit. med. J. **1969 II**, 662.
3. CROWTHER, D., BATEMAN, C. J. T., VARTAN, C. P., WHITEHOUSE, J. M. A., MALPAS, J. S., HAMILTON FAIRLEY, G., SCOTT, R. B.: Brit. med. J. **1970 IV**, 513.
4. MALPAS, J. S., SCOTT, R. B.: Brit. med. J. **1968 III**, 227.
5. MALPAS, J. S., SCOTT, R. B.: Lancet **1969 I**, 469.
6. MATHE, G., AMIEL, J. L., HAYAT, M., DEVASSAL, F., SCHWARTZENBERG, L., SCHNEIDER, M., JASMIN, C., ROSENFELD, C.: Presse méd. **78**, 1997 (1970).

Part VI

Short Communications

Adriamycin in Mesenchymal Tumors and Leukemias — Results of Two Years' Research

G. BILE, R. DE BIASI, and G. BERARDI

XV. Divisione Medica: Sezione Ematologia (Direttore: Prof. G. BILE) Servizio Trasfusionale e di Immunoematologia (Direttore: Prof. R. DE BIASI) Ospedale dei Pellegrini — Napoli

Materials and Methods

Adriamycin was given to 29 patients with Hodgkin's disease (8 cases), reticulosarcoma (4 cases), lymphosarcoma (3 cases), chronic lymphatic leukemia (7 cases), chronic myeloid anemia (2 cases), acute leukemia (5 cases); 20 patients were males, 9 females. Three further patients with other solid tumors are not considered, as well as 6 more patients inadequately treated.

Adriamycin was given at a daily dose of 0.35—0.5 mg/kg for 6 days and courses were repeated 2–7 times at intervals of 20–30 days. As evaluation criteria for blood diseases, we adopted those suggested by the Research Team on Chemotherapy (HEWLETT et al.), for the solid tumors we followed the categories suggested by by KARNOFSKY.

Toxicity

Among side effects, in our patients alopecia was particularly frequent (16 cases); in order of frequency followed: gastro-enteric troubles (mostly diarrhoea, of no high intensity and easily cured — 8 cases); stomatitis, mostly of the ulcero-necrotic type (5 cases); polyneuritis (1 case); and cutaneous rash (1 case).

One patient showed strong intolerance of treatment with severe shock and so the treatment was withdrawn.It is also to be noted that the same patient also showed similar manifestations of intolerance to other drugs (6-mercaptopurin and azotyprine). At no time were cardiotoxic manifestations observed, either clinically or electrocardiographically.

Nor were there significant alterations of haematochemical findings (e.g. BUN, uricacidemia, glycemia, alkaline phosphatase, transaminases, blood proteins). As far as blood cells are concerned, a certain leukopenia was always apparent, both when there was hypercytosis and when the number of blood cells was normal. In no case was thrombocytopenia observed.

Previous Therapies and Therapeutical Associations

In a large majority of the cases, adriamycin was used alone, except in a few cases in which, because of bad reactions with high fever, it was necessary to give steroids for some days; when the leukopenic effect was strongly felt and severe stomatitis occurred, we associated carbenicillin at a mean dose of 30 gr a day, by intravenous infusion, with good results. On acute leukemias transfusional therapy was also employed. As far as previous therapy is concerned, nearly all cases of malignant lymphogranuloma had already received different; treatment schedules, (procarbazine, vinblastine, chlorambucil, roentgen therapy and steroids.)

Results

Results are reported in Tables 1 und 2.

Hodgkin's disease, lymphosarcoma, reticuloblastoma and chronic lymphatic leukemia appeared the most sensitive to adriamycin; acute leukemia was less responsive and chronic myeloid leukemia only slightly. Regression in lymphomata lasted usually 2 or 3 months.

Table 1

Cases	Diagnosis	Dose in mg/kg/day	Courses No	Total dosage	Results C.R. P.R.	Toxicity
1	Acute haemocitoblastic leukemia	0.38	2			
2	Acute haemocitoblastic leukemia	0.35	2			
3	Acute haemocitoblastic leukemia	0.37	1			
4	Acute haemocitoblastic leukemia	0.47	1			
5	Acute monoblastic leukemia	0.28	1			
6	Chronic myeloid leukemia	0.35	2			
7	Chronic myeloid leukemia	0.36	1			
8	Chronic lymphatic leukemia	0.36	2			
9	Chronic lymphatic leukemia	0.30	1			
10	Chronic lymphatic leukemia	0.36	3			
11	Chronic lymphatic leukemia	0.36	2			
12	Chronic lymphatic leukemia	0.36	2			
13	Chronic lymphatic leukemia	0.25	2			
14	Chronic lymphatic leukemia	0.32	2			

Adriamycin at the doses used did not show great toxicity and in particular no cardiac toxicity found.

The drug appeared a good remission inducer especially in solid tumors, even if regression was no longer than that achieved with conventional therapy.

Table 2

Cases	Diagnosis	Dosage in mg/kg/day	No courses	Results according to KARNOFSKY	Total dosage	Toxicity
1	Hodgkin's disease	0.5				
2	Hodgkin's disease	0.35	1	I—B		
3	Hodgkin's disease	0.38	2	I—B		
4	Hodgkin's disease	0.32	4	I—A		
5	Hodgkin's disease	0.46	4	I—A		
6	Hodgkin's disease	0.43	3	I—B		
7	Hodgkin's disease	0.43	2	I—A		
8	Reticulosarcoma	0.33	1	I—A		
9	Reticulosarcoma	0.38	2	I—B		
10	Lymphoreticulo-sarcoma	0.30	4	I—B		
11	Lymphoreticulo-sarcoma	0.30	4	I—A		

Preliminary Clinical Observations on the Antitumor Effect of Adriamycin

G. Cardinali

Laboratorio di Ematologia-Istituto di Genetica Medica dell'Universita di Roma e Reparto di Radiologia, Ospedale S. Camillo Ospedali Riuniti di Roma.

With 3 Figures

Adriamycin (A.) is a new antibiotic with antitumor activity recently isolated from a mutant of *Streptomices Peucetius*, *S. Peucetius caesius*, in the Research Laboratories of Farmitalia. The structural formula of A. is very close to that of daunomycin which was isolated from *S. Peucetius* [1].

Experimental studies on animal tumors showed that, at equitoxic doses, A. has a greater antitumor activity than daunomycin [2, 3]. Preliminary clinical investigations demonstrated that A. is an active antitumor agent in man [4–8].

Material and Method

The following cases were treated in this investigation: acute myeloid leukemia 1, acute lymphatic leukemia 1, blastic transformation of chronic myeloid leukemia 3, and reticulum cell sarcoma 2.

Adriamycin (A.) was given i.v. at the single dose of 0.3–0.4 mg/kg of body weight, for 2–4 consecutive days. In almost all cases vincristine, 1 mg at weekly intervals, and prednisone, 40–70 mg/day, were associated. In four cases A. was continued, on a maintenance basis, for some weeks at the dosage of 0 mg once or twice a week.

Blood counts, serum electrolytes, serum electrophoresis, transaminases, ECG, bone marrow smear examination and routine blood testswere obtained for all patients before and at various intervals after starting treatment.

Results

Two patients, one with acute myeloid leukemia and one with blastic transformation of chronic myeloid leukemia were in a preterminal stage and, therefore, could not be adequately treated. The first of these 2 patients received only a 3-day course of A. and the second patient 4 administrations of the compound. Neither patient showed any apparent beneficial effect on the progressive course of the disease.

Good partial remission was obtained in a 51-year-old patient with acute lymphatic leukemia (Fig. 1) after two 4-day courses of A., following a short treatment with arabinosylcytosine. There was rapid fall in the white blood cell count, which, however, did not reach a normal level, and a progressive decrease in the percentage of blasts in the peripheral blood. A marked reduction of bone marrow blasts was also observed. A bone marrow examination, made at about $2^1/_2$ months after discontinuation of A., showed only 12% blasts and good erythropoietic activity.

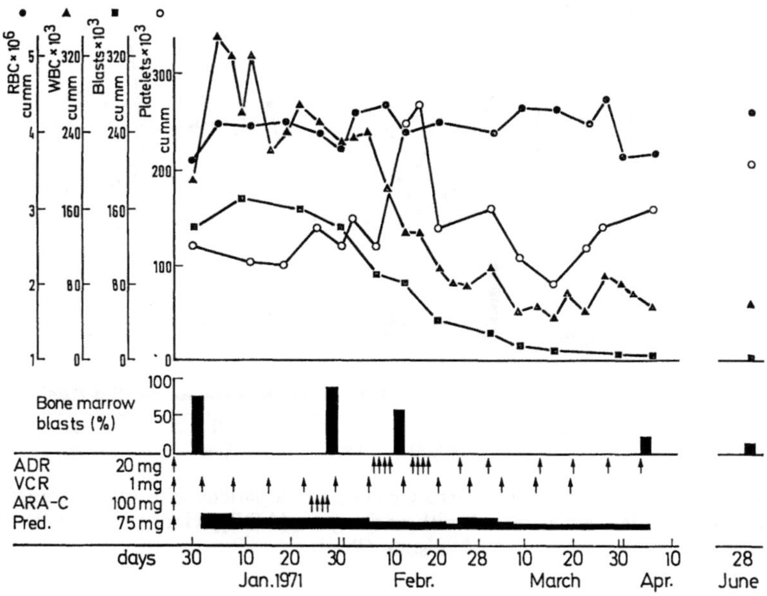

Fig. 1. Peripheral blood and bone marrow changes in a patient with acute lymphatic leukemia treated with adriamycin (ADR), vincristine (VCR), arabinosylcytosine (ARA-C) and prednisone (Pred)

Two patients in the acute phase of chronic myeloid leukemia responded favourably to A. One of them, a 47-year-old patient (Fig. 2) had, after a 3-day course of A., a rapid fall in the white blood cell count with disappearance of blasts from the peripheral blood. Blasts, however, reappeared in the blood after a few days, in spite of a progressive decrease in the bone marrow blasts. The lowest level reached by bone marrow blasts was 12%. After about 3 weeks the bone marrow showed a marked increase in the percetage of blasts, which also greatly increased in the peripheral blood. The patient did not respond to further treatment with A. There was a profound fall in red blood cell and platelet counts and the patient eventually died with symptoms of gastro-intestinal hemorrhage.

The second patient with acute transformation of chronic myeloid leukemia received four 4-day courses of A. to a total of 320 mg. There was only a moderate reduction in the white blood cell count and a decrease of peripheral blasts from 30 to 8%. The bone marrow showed only a moderate decrease in the percentage of blasts. Fever disappeared after the first 3 days of treatment and the size of the spleen showed a 25% reduction. It must be pointed out that this patient had been previously treated with various agents including myleran, 6-mercaptopurine. hydroxyurea, trenimon, cyclophosphamide, daunomycin, vincristine, prednisone, arabinosyl-

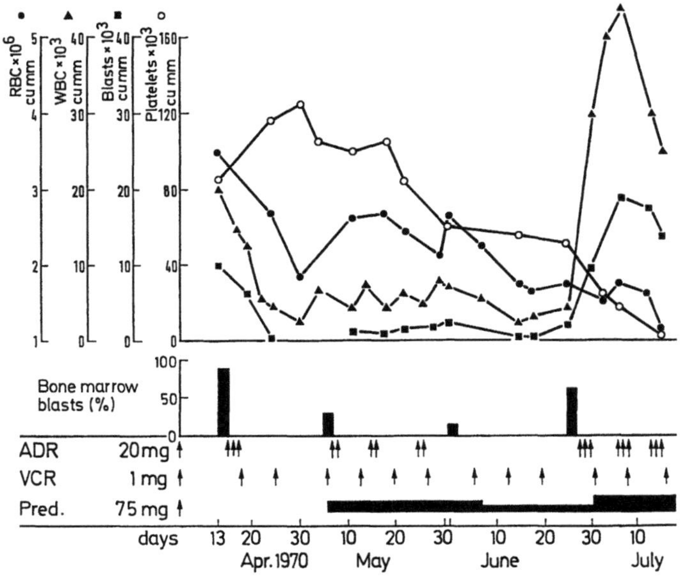

Fig. 2. Peripheral blood and bone marrow changes in a patient with blastic transformation of chronic myeloid leukemia treated with adriamycin (ADR), vincristine (VCR) and prednisone (Pred)

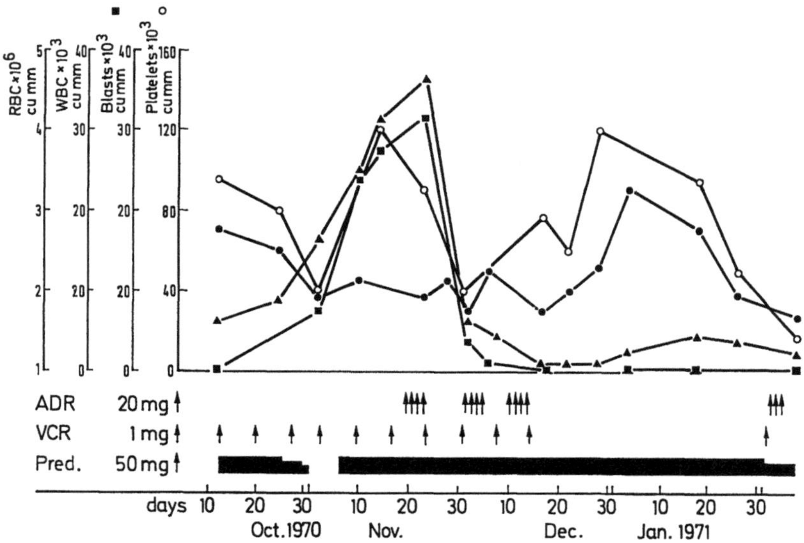

Fig. 3. Peripheral blood changes in a patient with reticulum cell sarcoma treated with adriamycin (ADR), vincristine (VCR) and prednisone (Pred)

cytosine and had become resistant to all of them, with the possible exception of arabinosylcytosine. This patient was treated again with A. about 3 months later and then kept on continuous treatment on a maintenance basis. He never obtained a true remission but he again showed some improvement both in the hematological and clinical picture. This patient is still alive more than 15 months after the conversion of the disease.

Good partial remissions were obtained in both patients with reticulum cell sarcoma. Treatment with A. was followed by general improvement of the clinical picture, more than 50% regression of lymph node enlargements and other tumor manifestations, and disappearance of fever.

In one of the two cases the course of the disease was complicated (Fig. 3) by a marked increase in the white blood cell count with the appearance in the peripheral blood of a high percentage of blastic cells of lymphoreticular type. The white blood cell count returned to normal and the blasts disappeared after the treatment. The partial remission lasted 6 weeks. In the other patient with reticulum cell sarcoma the remission could be maintained, with continuous treatment, for about 8 months.

Summary

Good partial remissions were obtained in one case of acute lymphatic leukemia, one case of blastic transformation of chronic myeloid leukemia and in 2 cases of reticulum cell sarcoma. In the other case of blastic transformation of chronic myeloid leukemia a prolonged improvement of the clinical and hematological picture was obtained.

Toxic effects consisted mainly in bone marrow depression and alopecia, which was constantly observed in the patients receiving prolonged treatment. Ulceration of oral mucosa was seen in 2 patients. Nausea occurred in 2 patients. No ECG changes were observed.

References

1. Grein, A., Spalla, C., DiMarco, A., Canevazzi, G.: Gion. Microbiol. **11**, 109 (1963).
2. DiMarco, A., Gaetani, M., Scarpinato, B.: Cancer Chemother. Rep. **53**, 33 (1969).
3. Sandberg, J. S., Howsden, F. L., DiMarco, A., Goldin, A.: Cancer Chemother. Rep. **54**, 1 (1970).
4. Bonadonna, G., Monfardini, S., DeLena, M., Fossati-Bellani, F.: Brit. med. J. **1969 III**, 503.
5. Monfardini, S., Bonadonna, G., DiPietro, S., Giuandani, A., Fossati-Bellani, F., DeLena, M.: Tumori **55**, 197 (1969).
6. Massimo, L., Cottafava, F., Mori, P. G., Fossati-Guglielmoni, A.: Minerva Pediat. **21**, 2182 (1969).
7. Bonadonna, G., Monfardini, S., DeLena, M., Fossati-Bellani, F., Beretta, G.: Cancer Res. **30**, 2572 (1970).
8. Mathe, G., Amiel, J. L. Hayat, M., DeVassal, F., Schwarzenberg, L., Schneider, M. Jasmin, Cl., Rosenfeld, Cl.: Presse med. **78**, 1997 (1970).

Preliminary Clinical Trial of Adriamycin in Acute Leukemia and Solid Tumors of Children

AYHAN O. ÇAVDAR, Ü. TOPUZ, E. BABACAN, A. ARCASOY, and S. GÖZDAŞOĞLU

Department of Pediatrics (Hematology Section) Medical School of Ankara University
Ankara, Turkey

Adriamycin[1] was used on 10 children with acute leukemia and malignant tumors (7 acute leukemia and 3 solid tumors). The malignant tumors were rhabdomyosarcoma and neuroblastoma. The type of leukemia was ALL in 6 cases and AML in one. None of these 10 patients had ever received adriamycin; 4 of the 7 children with acute leukemia had previously been treated with other chemotherapeutic agents and 3 were new cases. Ages ranged from 1.5—12 years and there were 7 boys and 3 girls in the group.

Complete blood counts, bone marrow aspiration, liver function tests, NPN, serum uric acid, ECG and electrolytes were tested prior to therapy. Chest X-rays and skeletal surveys were also obtained before the adriamycin treatment. Blood counts were repeated three times weekly during therapy.

Dosage

Adriamycin was given by a rapid single IV injection at the dosage of 0.5 mg/kg daily for 4 consecutive days with 3-days interval, then 2 more doses of adriamycin depending upon WBC and platelets counts. A total of 2—3 mg/kg per course was used for acute leukemia. The course was repeated in 2 to 4 weeks in some children. Two leukemic children were given vincristine and steroid in addition to adriamycin. Two dose schedules were used for malignant tumors:

a) 0.4 mg/kg/IV on alternate days for 8 injections.

b) 0.5 mg/IV daily for 6—8 doses.

Results

2 children with ALL were lost at an early phase of the therapy whereas the third leukemic child had to be discharged immediately after the first course of adriamycin. Therefore these 3 patients could not be evaluated properly. 3 of the 4 remaining children with acute leukemia had remissions (one complete and 2 good partial remissions).

1 Kindly supplied by Farmitalia

The patient with AML did not respond to adriamycin, despite the fact that the drug was used in combination with vincristine and prednisone.

The case with embryonal rhabdomyosarcoma, initially thought to be neuroblastoma, showed objective and suvjective improvement, the tumor grew smaller but did not disappear. Two children with neuroblastoma, one being fresh and not the metastatic abdominal type, received two courses of adriamycin plus cobalt teletherapy with only mild to moderate response. The second neuroblastoma patient had generalized bone metastases and previous chemotherapy (VCR + End.). He was given adriamycin recently and showed only a subjective improvement, namely decrease of bone pains.

As to the side effects, alopecia has been the most common one. Oral ulcers leukopenia and fever were observed in 5 children, Bone marrow depression was usually seen at the end of the second week of therapy. Diarrhea was observed in two patients. No significant changes in ECG, and, blood chemistry have been noticed.

Although the number of cases were not enought to justify a definite conclusion, we had the impression adriamycin could be a useful drug particularly in leukemia.

For the time being, we are unable to evaluate the duration of remission as the time is too short.

Combined Chemotherapy of Acute Leukemias

H. Gerhartz

Hemat.-oncol. Department (Klinikum Westend), Free University Berlin

For induction and reinduction therapy in patients with acute leukemias, we used daunomycin 60 and adriamycin in 19 adults, mostly in combination with vincristine and prednisone (Table 1). Furthermore, 18 patients were treated with the 4-drug combination of asparaginase, daunomycin, vincristine and prednisone, and 30 patients with citosine arabinoside and prednisone. The duration of one course of treatment

Table 1. Combined treatment in acute leukemias

	Pat.	Remiss. %	Mean durat. of courses days	Mean total dose mg	Mean dose per day mg
C-Arab.	30	33	9	814	90
+ Predn.				1250	140
Daunomycin	60	47	18	286	16
+ Vincr.				7.1	0.42
+ Predn.				1450	82
Adriamycin	19	47	13	186	14
+ Vincr.				4.8	0.37
+ Predn.				1170	90
Asparaginase	18	50	18	127000 I.E.	7500 I.E.
+ Daunomycin				190	11
+ Vincr.				4.6	0.25
+ Predn.				1460	81

extended over 2—4 weeks, averaging 9 days with C.-Arab. and over 18 days with daunomycin. With adriamycin the mean duration extended to 13 days, a shorter period when stomatitis did occur. The mean total dose per course amounted to 286 mg for daunomycin and 186 mg for adriamycin. The dose ratio between adriamycin and daunomycin was virtually 2 : 3. In combination with asparaginase, only two-thirds of the usual dose of daunomycin and vincristine could be administered.

In acute leukemias with equitoxic doses, adriamycin achieved the same rate of remissions as daunomycin (47% in adults). Adriamycin had slightly less influence on the normal hemopoiesis and decreased the leucocytes and thrombocytes more slowly and less intensively than daunomycin. Cardiotoxic disturbances were observed less frequently after adriamycin, but adriamycin causes more severe stomatitis and esophagitis and a greater loss of hair. The various forms of acute leukemia respond differently to chemotherapy. Remissions are achieved in adults only approximately in half the number and are slightly incomplete. But as the differentiation of the various forms is difficult at times, it is permissible to state the total number of all cases with regard to the clinical results first. The remission rate reached 44% for cytostatics in at least 109 patients and 21% for the enzyme therapy with asparaginase in 181 patients, although the first group consisted nearly exclusively of adults and one-third of the latter group were children. In acute non-lymphoblastic leukemias, 41% of remissions were found with cytostatics in 91 patients, and 8% with asparaginase in 96 patients.

The highest rate of remissions was found in lymphoblastic leukemias and these declined gradually in the order: promyelocytic leukemias, the myeloblastic crisis of chronic myelogenous leukemias, monocytic and myeloblastic leukemias. In Table 2 the first figure gives the number of patients and the second figure the number of remissions with the different drugs and leukemic forms.

Table 2. Number of patients-number of remissions

	ALL	AML	AMOL	APL	CMLM
C-Arab.		19— 6	4—1	4—1	2—2
D + VC + P	9— 7	28—10	3—2	11—5	4—2
A + VC + P	2— 2	11— 4	1—0	3—2	2—1
Asparaginase	85—30	65— 5	8—0	16—4	9—0
Asp. + D + VC + P	12— 7	4—1	1—1		1—0
Total pat.	108	123	17	34	18
% remissions	43	21	23	35	28

Table 3

	ALL		AML		AMOL		APL		CMLM	
	pat.	% remiss.	pat.	% remiss.	pat.	% remiss.	pat.	% remiss.	pat.	% remiss.
C-Arab. + P			19	32	4	25	4	25	2	100
D or A + VC + P	11	82	38	37	5	40	14	50	6	50
Total			57	35	9	33	18	44	8	62
Asparaginase	85	35	61	8	8	0	16	25	9	0
Asparaginase +D+VC+P	12	58	4	25	1	+			1	0

In order to simplify this division into many groups and to make comparison easier, we have given the remissions as percentages in spite of the fact that we had fewer cases in some groups (Table 3). Daunomycin and adriamycin have been gathered into one group. In the lymphoblastic leukemias (ALL) remissions were seen in 82% with the combination of daunomycin (D) or adriamycin (A) with vincristine (VC) and prednisone (P), and in 35% when asparaginase was used as monotherapy. Giving asparaginase simultaneously in comparison with the 3-drug combination did not improve the remission rate and less was achieved than when the 3 drugs were used on their own.

With the combination of daunomycin or adriamycin, vincristine and prednisone, the remission rate was 50% in the promyelocytic leukemias (APL) or myeloblastic crisis (CMLM), 40% in monocytic (AMOL) and 37% in myeloblastic (AML) leukemias. The rate was a little higher than with C.-Arab. It may be, however, that C.-Arab. is superior in myeloblastic crisis.

The monotherapy of C-Arab. and also of daunomycin or adriamycin reached with 50% and 54% distinctly better remissions than the combination of each drug with prednisone (27—47%) or the 3-drug combination (45%). This leads us to the conclusion that combination therapy is not to be considered basically and at all times as the best, particularly in the case of the combination with prednisone and in the non-lymphoblastic forms; conversely, in lymphoblastic leukemias combination therapy seems to be superior to monotherapy. Within the range of acute leukemias, the various forms are influenced differently in extent and according to the different drugs or drug combinations.

Clinical Trial of Adriamycin on Malignant Diseases

YOSHIO HARA and YUZO KURITA

Cancer Center, Niigata Hospital, Niigata, Japan

We have used adriamycin on 27 cases since March '71 and have had good results as shown in Table 1, especially in the cases of lung cancer.

Our Dosage Schedule

0.4 mg/kg every other day × five times, then once a week.

Criteria of Clinical Effects

We judged clinical effects by KARNOFSKY's criteria.

Clinical Results

1) Lung cancer: out of 18 cases, three showed O—B, three showed O—C, one 1—A, and two showed 1—B.
2) Gastric cancer: out of four cases, two showed O—B.
3) Pancreas cancer: one out of two showed 1—B.
4) Hodgkin's disease: one showed O—O.
5) Lymphosarcoma: one showed O—O.
6) Reticulosarcoma: one showed 1—B.

Table 1. Clinical result of adriamycin treatment

	No. of cases	O–O	O–A	O–B	O–C	1–A	1–B	1–C
Lung cancer	18	9		3	3	1	2	–
Gastric cancer	4	2		2				–
Pancreas cancer	2	1					1	–
Hodgkin's disease	1	1						–
Lymphosarcoma	1	1						–
Reticulisarcoma	1						1	–
Total (by KARNOFSKY's criteria)	27	14		5	3	1	4	–

Side Effects

1) Bone marrow suppression: 2 out of 19 cases showed leucopenia (less than 3000) and two out of 20 cases showed thrombocytopenia (less than 100000).

2) Alopecia: alopecia was seen in 12 out of 27 cases.

3) Stomatitis: stomatitis was seen in nine out of 27 cases.

4) Anorexia: 8 out of 27 cases complained of loss of appetite.

5) ECG changes: 4 out of 15 showed flat-T; 2 showed depressed-ST. One showed auricular fibrillation and one showed sinus bradycardia.

6) Electrolyte balance: None showed significant change. We would like to present several cases who showed good results by adriamycin.

Adriamycin in the Chemotherapy of 21 Solid Tumors
Preliminary Results

Piero Lovisetto and Giuseppe Pochettino

Clinica Medica Università di Torino (Direttore: Prof. G. C. Dogliotti)

With 2 Figures

The following preliminary results with adriamycin hydrochloride were obtained in 21 solid tumors (Table 1). From two to four 3-day 0.5 mg/kg/d cycles spaced at 7 (or more) day intervals were administered. The drug was dissolved in distilled water immediately before i.v. injection over 2—3 min via a venous drip tube, followed by 250 cc saline to wash the vein.

Clinical Results

A. Epithelial Tumors

17 cases, including: 1 lymph node metastasis secondary to cancer of the penis, 1 bladder cancer, 3 lung cancer, 2 melanoma metastasis, 4 gastroenteric tract

Table 1. Karnofsky response in 21 cases treated with adriamycin

Diagnosis	No. of cases	Karnofsky response category O–O	O–A	O–B	O–C	I–A	I–B	I–C	II
Penile cancer	1					1			
Bladder cancer	1								1
Lymphosarcoma	2			1			1		
Rhinopharyngeal cancer	1				1				
Lung cancer	3	1		1	1				
Melanoma	2				2				
Gastrointestinal cancer	4	1		1					2
Pancreas cancer	1	1							
Hypernephroma	1	1							
Breast cancer	3	1							2
Osteogenetic sarcoma	1	1							
Skin and multiple visceral metastases secondary to cancer of unknown origin	1	1							

neoplasia, 1 pancreas cancer, 3 breast and 1 rhinopharynx with multiple metastases, 1 skin and multiple visceral metastases secondary to cancer of unknown origin. For clinical and laboratory data, see Tables 1 and 2.

Table 2. ECG findings in 17 patients examined systematically before during and after treatment

ECG before treatment	No. of cases	ECG after treatment	No. of cases
Within the limits of normal	10	Unchanged	5
		Onset of abnormalities	5
Various forms of abnormality	7	Unchanged	6
		Improved	1
		Deteriorated	—

B. Other Types

4 cases, including: 1 hypernephroma with widespread lung, multiple subcutaneous, and hepatic metastases; 1 lung metastasis secondary to osteosarcoma; 1 intestinal lymphosarcoma; and 1 widespread lymphosarcomatosis. The results are shown in Tables 1 and 2.

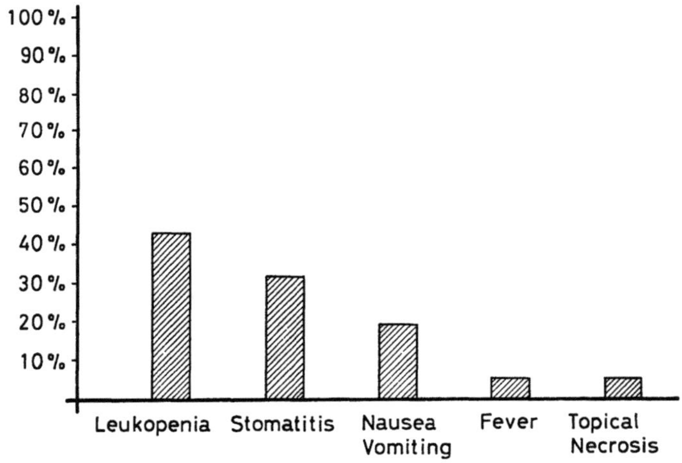

Fig. 1. First toxic side effect

Toxicity and Unwanted Side-Effects

Toxic signs were a constant feature and sometimes compelled suspension of the treatment. The first to appear (Fig. 1) were leukopenia (43.7% of cases) and stomatitis (31.2%). The end-of-treatment picture (Fig. 2) almost always included leukopenia (100% of cases), stomatitis (81%) and loss of hair (75%). White cell decreases ranged from 40—90% and were apparent 10—12 days (or even 6—7 days) after the commencement of the treatment. Return to normal was usually observed 7—8 days after suspension. There were no instances of irreversible marrow aplasia.

Stomatitis began as burning and redness of the mucosa. After 2—3 days, ulcers appeared on the edges of the tongue and around the commissurae labiorum.

Alopecia was a virtually constant sign in patients who completed their course of treatment. Regrowth was observed within 3—4 months after suspension, as reported by BONADONNA et al. (1970).

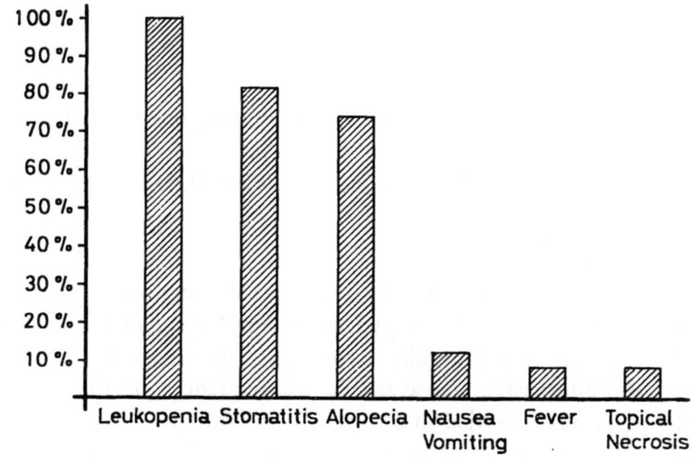

Fig. 2. Toxic side effects appeared during treatment

Nausea and vomiting (19% of cases) appeared immediately after the first cycle, or even after the first administration. The cardiotoxicity of the drug was assessed in the light of the results of systematic ECG examination before, during and after treatment in 17 subjects (Table 2).

No significant changes were noted in blood nitrogen, transaminase, phosphatase, uric acid or total protein values.

Conclusions

Adriamycin inhibits the synthesis of nucleic acids, particularly DNA, by attaching itself to chromatin in the premitotic stage. It is thus of considerable theoretical interest as an antiblastic, and has proved very effective in the experimental animal (DI MARCO et al., 1969). In our opinion, however, these theoretical possibilities cannot be said to have borne fruit under clinical conditions.

Our series was confined to solid tumors, whose resistance to all forms of chemotherapy is well known. Adriamycin proved completely ineffective in 25%, and in a further 25% nothing more was observed, than a slowing down of the course of the disease, unaccompanied by distinct signs of improvement. Of the remaining cases in which regression — never longer than 1 month — was noted, only two (1 lymph node metastasis secondary to cancer of the penis; 1 intestinal lymphosarcoma) displayed clearly defined and significant improvement.

The toxicity of the drug, particularly with respect to the heart, may be considered as an open question for the moment, through its lesser cardiotoxicity by comparison with daunomycin should be borne in mind.

Adriamycin Use in Solid Tumors and in Lymphoblastic Leukemia of Children

E. Madon, P. G. Baroncelli, A. Sacerdote, and P. L. Roberi

Clinica Pediatrica dell'Università di Torino (Direttore: M. Gomirato Sandrucci)

Adriamycin has been used für the past 3 years in treating solid tumors and acute lymphoblastic leukemia in the "Oncology Department" of the Pediatric Clinic, Turin.

The drug[1] was administered initially to subjects in the terminal stage of the disease for whom no valid alternative was available. Subsequently, on the basis of the experience achieved, it was included in our polychemotherapy schedules.

In these 3 years, 19 patients, whose ages ranged from 18 months to 11 years, were treated. In particular, there were: 8 patients affected with neuroblastoma, 3 with malignant lymphogranuloma, 3 with acute lymphoblastic leukemia, 2 with osterosarcoma, 1 with retinoblastoma, 1 with testicular teratoma and 1 with granulomatous thymoma.

We have followed the treatment scheme suggested by the manufacturer, administering adriamycin in cycles of 3 consecutive days at the dose of 0.5 mg/kg/d with an interval of 7 days. Recently, we have changed the dose pro kg/d to 0.6 mg and the interval to 10 days.

There were normally 3 cycles, followed by a single weekly administration at the same dose, till the appearance of toxic signs or remission. It was not possible or useful in all cases to persist with this therapeutic scheme, either because of serious toxic effects, or because there was no positive response.

Among the treated patients, we have excluded 3 from this report (2 cases of acute lymphoblastic leukemia and one of neuroblastoma) which were not useful for the assessment of the drug activity because they were treated in the terminal stage as a "last resort".

The results obtained in the remaining 16 cases are reported in Table 1.

It should be stressed that the drug was never administered as first-choise chemotherapy, but always in the relapse stage.

The toxic effects detected during the treatment or soon after, and in any case strictly related to the drug action, are reported in Table 2.

It should be emphasized that an exact evaluation of such effects is not always possible owing to the interference of the action of drugs previously administered. In fact, the alopecia observed in 10 out of 16 patients was already present in 4 patients before

1 Kindly supplied by Farmitalia.

Table 1. Response to adriamycin therapy. Evaluation according to KARNOFSKY's parameters

Type of neoplasia	No. of cases	Type of response O–O	O–A	O–B	O–C	I–A	I–B	I–C
Neuroblastoma	7	1	2	–	–	2	2	–
Malignant lymphogranuloma	3	–	–	–	–	–	3	–
Osteosarcoma	2	1	–	–	1	–	–	–
Retinoblastoma	1	–	–	–	–	–	1	–
Testicular teratoma	1	–	–	–	–	–	1	–
Granulomatous thymoma	1	–	–	–	–	–	1	–
Acute lymphoblastic leukemia	1	Total remission for 2 months						
Total	16	2	2	0	1	2	8	0

Table 2. Evaluation of adriamycin toxicity on 16 treated patients

Alopecia	10
Leukopenia	8
Stomatitis	6
Hyperpyrexia	4
Nausea-vomiting	6
Diarrhea	6
Hemorragic syndrome	3

treatment with adriamycin, due to the toxic effect of vincristine, previously adminstered.

Lastly it is important to stress that the first toxic sign noted was: vomiting in 4 cases, leucopenia in 6 cases and stomatitis of a severe type in 5 cases.

In conclusion, owing to the heterogeneity and the small number of patients, it is impossible to draw valid conclusions from our case list.

We shall limit ourselves to pointing out the results obtained, which may be considered positive in 70% of cases. These data will encourage us to continue the clinical trials of this drug in the therapy of childhood neoplasias.

Preliminary Data on the Treatment of Hemoblastoses with Adriamycin

G. Marinone, E. Maiolini, and E. Spadacci

II. Divisione Medica dell' Ospedale Civile di Udine

With 8 Figures

13 adults and 2 children (Table 1) suffering from leukoses (5 cases) or from systemic lymphopathies (10 cases), 9 already pretreated and 6 at first stage of the disease, were treated with adriamycin at the dose of 0.30—0.40 mg/kg/d for 3 consecutive or alternating days for 2—4 cycles with an interval of 4—10 days in between.

Table 1

Diagnosis	Adults	Children	No previous treatment	Total
Acute leukemia	3	2	2	5
Lymphosarcoma	5	—	2	5
Stem cell crises in myeloid leukemias	2	—	—	2
Hodgkin's disease	2	—	1	2
Multiple myeloma	1	—	1	1
Total	13	2	6	15

Table 2. Acute leukemias

Diagnosis	Drug	Dosage mg/kg	Days	Disease durat. in months	Remission Type	Remission in days
ALL	Adriamycin	0.3—0.4	1—2—3	7	Partial	—
	Prednisone	1.5—2.0	Every day			
ALL	Adriamycin	0.3—0.4	1—2—3	22	Complete	70
	Prednisone	1.5—2.0	Every day			
ALL	Adriamycin	0.3—0.4	1—2—3	22	Partial	40
	Prednisone	1.5—2.0	Every day			
AML	Adriamycin	0.3—0.4	1—2—3	—	Partial	22
	Prednisone	1.5—2.0	Every day			
A. Histoblastic leukemia	Adriamycin	0.3—0.4	1—2—3	—	Partial	43
	Prednisone	1.5—2.0	Every day			

Adriamycin has been given alone in some cases, in others (leukemias) associated with prednisone (at the dose of 1.5—2 mg/kg/d, daily) or Hodgkin's disease and lymphosarcoma with prednisone (1 mg/kg/d, daily) and vincristine (0.05 mg/kg for one day).

Clinical Results

Results obtained in the treatment of acute leukemias with adriamycin and prednisone are reported in Table 2 and Fig. 1.

In this group, a child (Rosanna D, 15) suffering from acute histioblastic leukemia at onset was treated first with daunorubicine and prednisone without much effect (Fig. 2). Two cycles of adriamycin, administered immediately afterwards, brought

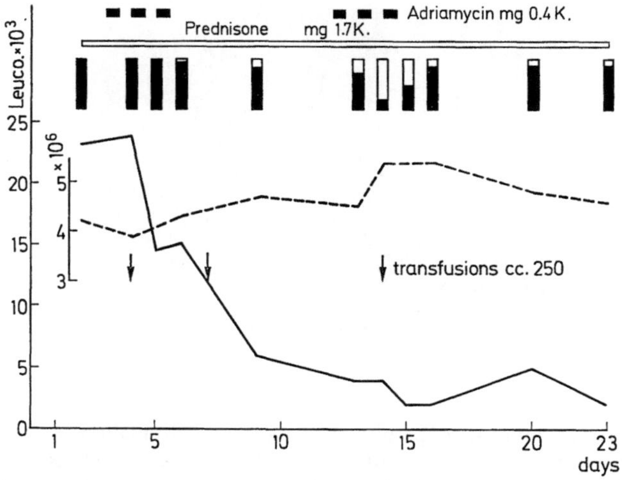

Fig. 1. Modifications of the leukocytes count in the blood of a child suffering from acute lymphoblastic leukemia and treated with prednisone and two cycles of adriamycin

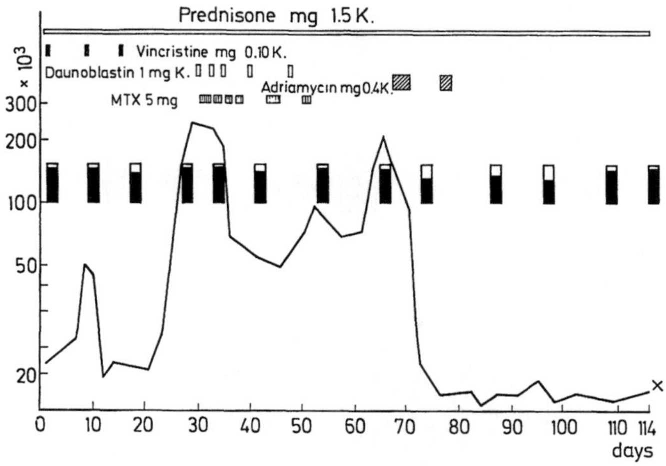

Fig. 2. Two cycles of adriamycin have induced normalization of the leukocyte count in a case of acute monocytic leukemia in which daunoblastine had induced a slight remission

the blood level of leukocytes back to normal with clear, though incomplete, improvement of the leukocyte differential count and of the bone-marrow composition.

Adriamycin proved to be active where daunorubicine failed. Also the *blastic crises of chronic myeloses* responded favourably to adriamycin in association with prednisone or prednisone and 6MP and a good partial remission was observed in both cases.

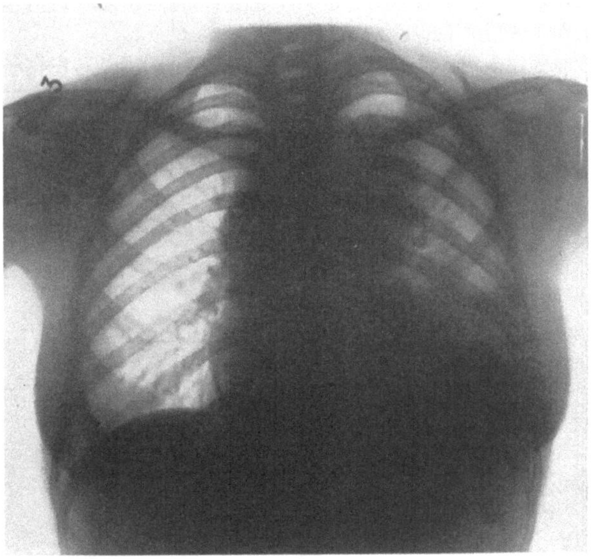

Fig. 3

Figs. 3, 4, 5. Mediastinum adenopathies in a patient with malignant lymphogranuloma slightly sensitive to X-ray treatment, after two adriamycin cycles (Figs. 3 and 4). Limited worsening of the radiologic picture twenty days after the second cycle (Fig. 5)

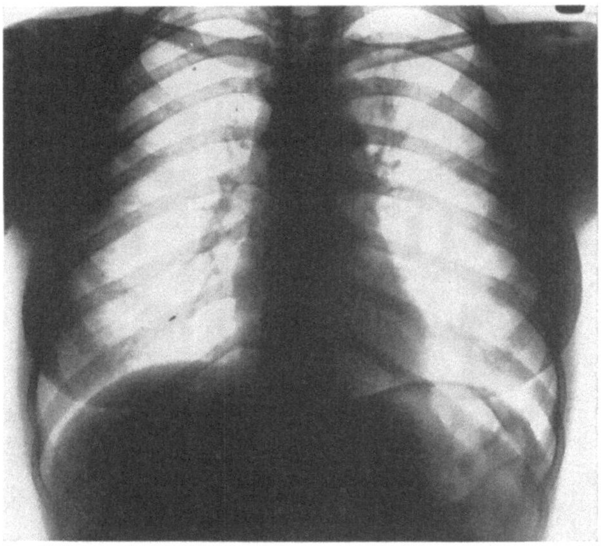

Fig. 4

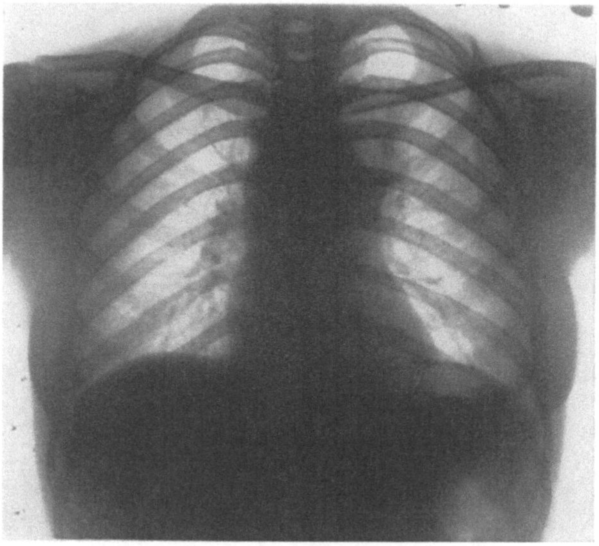

Fig. 5

Lymphosarcomas showed little benefit from the therapy, not so much from a qualitative point of view, but rather because improvement was brief: 4/5 complete clinical remissions, 2 complete medullary remissions and 3 complete hematological remissions were recorded. In no case, however, did remissions last more than 30 days. Relapses were always characterized by the reappearance of adenopathies. None of these patients was submitted to further treatment with adriamycin.

Two cases of Hodgkin's disease at the 3rd and 4th stages, respectively, were treated with the association vincristine-adriamycin-prednisone. The first patient had been pretreated for two months. In both cases, a few doses of adriamycin caused very quick and complete remission (Fig. 3, 4, 5) of the objective symptoms with disappearance of all adenopathies and of splenomegaly. The remission lasted only one month, so that at present we are repeating the treatment courses (vincristine-adriamycin-prednisone) every 20 days, in an attempt to stabilize the successful remissions obtained with the first course of treatment.

Toxic Activity

The toxic activity of adriamycin does not seem to be substantially different from that of daunorubicine, i.e. a *lytic action* on the undifferentiated stem cells in acute leukemias.

Such lytic action affects both the cytoplasm, which appears vacuolized (Fig. 6) and the nucleus which, besides macroscopic leaks of substance, often undergoes phenomena of more or less macroscopic fragmentation and segmentation and of structural homogenization. In the most serious cases the modified elements appear on the smear like "wood-worm eaten cells".

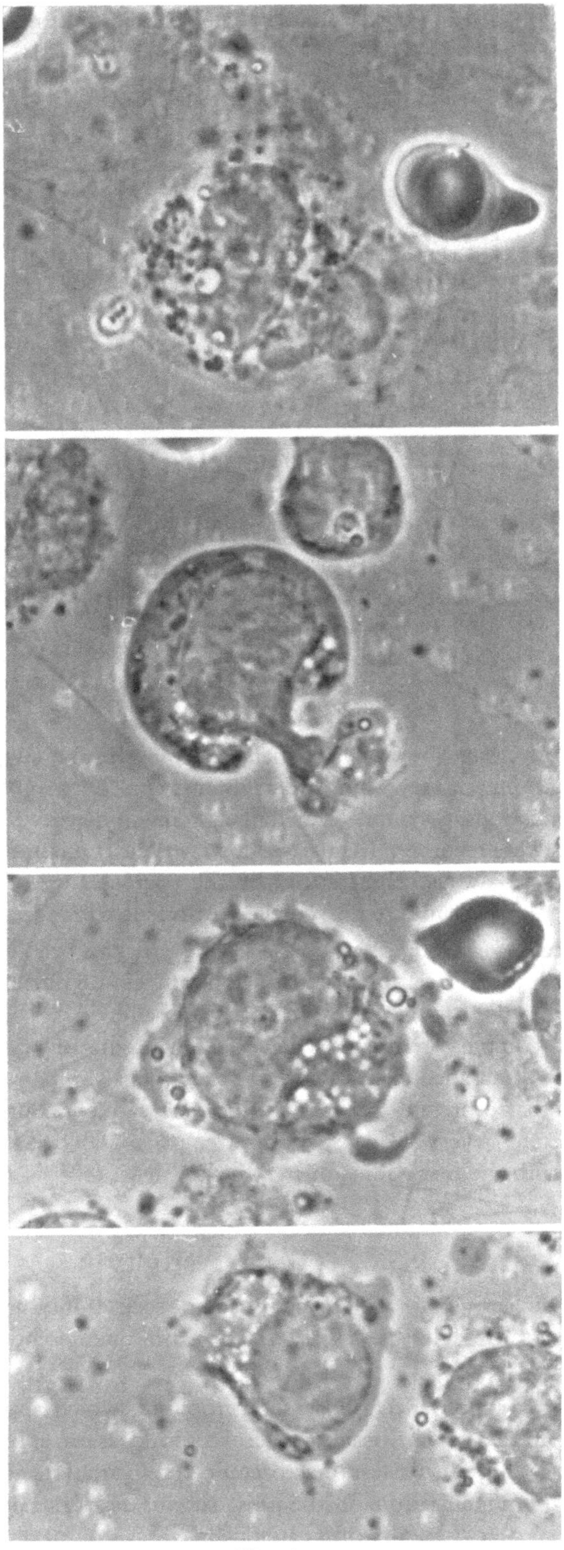

Fig. 6

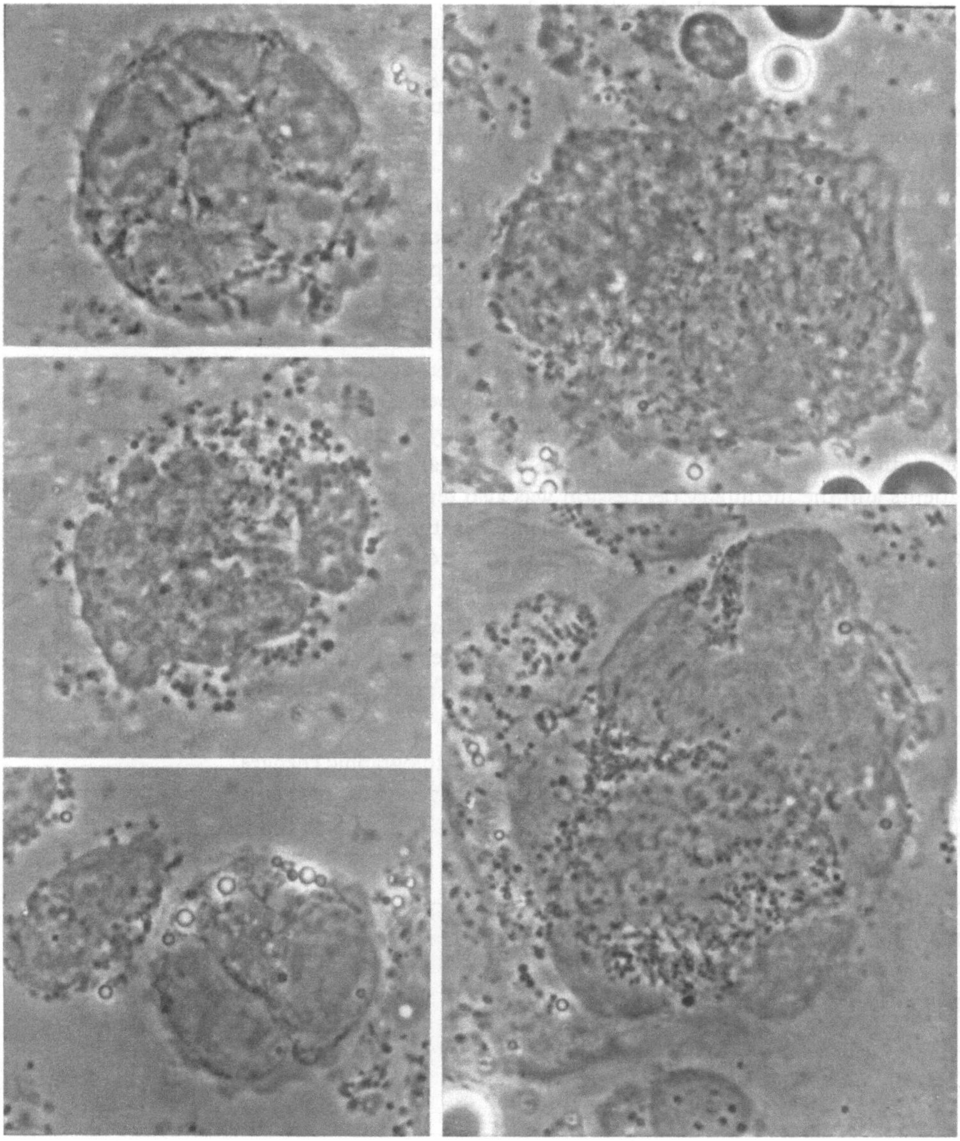

Fig. 7. Different degrees of cellular gigantism (megamyeloid cells) in bone marrow of acute leukemics treated with adriamycin (observation in wet preparation in phase contrast)

Few giant cells, like those observed with daunorubicine, have been observed (Fig. 7), whereas some megaloblast-like erythroblasts have frequently, been found in the bone marrow of A.L. patients, presumably due to action of the drug on the nucleic acid metabolism of bone-marrow cells (also reported by Bonadonna et al.). Rare orthochromatic megaloblasts and rare megalocytes have been also seen.

Fig. 6. Various degrees of cytoplasmic vacuolization in stem cells after adriamycin treatment (bone marrow observation on wet preparation in phase contrast)

In the two cases of blastic crisis of chronic myelosis treated with adriamycin, an impressive hyperplasia of basophilic granuloblasts at all maturation stages, was observed at the beginning of remission, suggesting that general basocytic hyperplasia (cellular gigantism and upsetting of the nuclear chromatinic structure were seen also in basophilic granuloblasts) may be a cytological warning sign of impending remission.

The second observation was suggested by one of the cases where myelosis had struck the carrier of a complete Pelger-Huet anomaly; the blastic modification had concealed the granulocytic anomaly, which reappeared after treatment with adriamycin. The drug was unable to affect the nuclear anomaly of the new granuloblastic generation, successive to the destruction of stem cells.

The *toxic* side effects of adriamycin (BONADONNA et al., 1969—1971; MASSIMO et al., 1969; WANG et al., 1971) are represented by alopecia, ulcerations of the oral mucosa and medullary depression. With a dose of adriamycin of 0.4 mg/kg/d clinical signs of the "mitotic arrest syndrome" (MARINONE et al.) were seldom observed in our patients and the medullary depression was also temporary; the loss of hair never got to the stage of true alopecia and the changes in the oral mucosa (burning sensation accompained by redness and swelling and, exceptionally, erosion) appeared only occasionally.

The cardiotoxicity of adriamycin seemed lower than that of daunorubicine: in 23 patients treated[1], 15 of whom were over 40, we never did observe any clinical sign of cardiopathies. ECG changes are reported in Table 3 and Fig. 8.

Table 3. Adriamycin — Electrocardiographic changes

Before treatment				After treatment		
Age	No. of cases	Normal	With minimal ECG abnormalities	Unchanged	Improved	Worse
Under 40	8	6	2	5	1	2 25%
Over 40	15	11	4	7 75% 53.3%	1	7 46.7%
Total	23	17	6	12 60.8%	2	9 39.2%

From the data reported above, it seems that even these preliminary studies may enable us to outline the prospects and limitations of adriamycin in the treatment of acute leukoses and malignant productive lymphopathies. The prospects are:

a) the treatment of *generalized malignant lymphogranuloma* where prompt and very good results can be obtained;

b) the treatment of *lymphoblastic leukoses resistant or unresponsive* to vincristine or daunorubicine;

1 (Including patients with tumors to whon we do not refer here.)

c) the treatment of *newly detected acute myeloblastic and monocytic leukoses, or of those in which a remission can no longer be induced with daunorubicine, methotrexate etc.*

The limitations are:

a) the *toxicity of the substance*, similar to that of daunorubicine, but with a lower cardiotoxicity;

b) the *shortness of remissions induced* (3—6 weeks, according to our experience and to the experience of most other researchers) adriamycin should therefore be regarded, at the most, as a mere "inducer" of remissions, as is the case for vincristine and daunorubicine;

c) the *high incidence of incomplete remissions.*

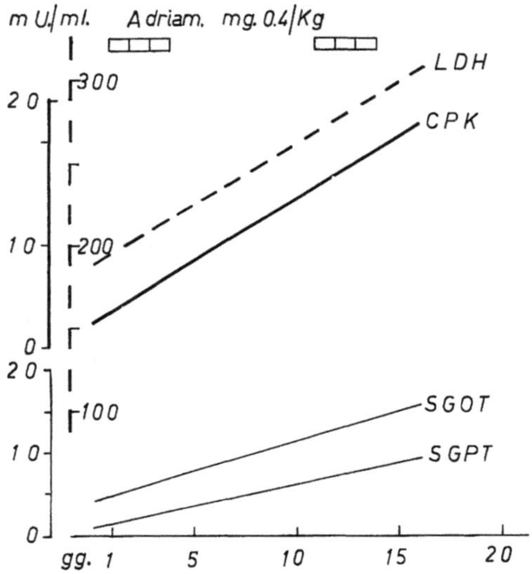

Fig. 8. Transaminase rate (SGOT, SGPT), creatinphosphokinase (CPK) and lactic dehydrogenase (LDH) in the blood of a patient treated with adriamycin, whose ECG was normal and who had even shown an improvement of the repolarization phase

Clinical Trials with Adriamycin

K. OTA

Aichi Cancer Center Nagoya, Japan

With 3 Figures

We report the response to and toxicity of adriamycin over a five-month period.

Method and Materials

16 patients with solid tumors (5 cases of lung cancer, 2 cases of reticulum cell sarcoma, one case each of carcinoma of larynx, stomach, rectum, ovary, cervix, and chondroblastoma, retroperitoneal sarcoma, malignant neurilemmoma and mediastinal tumor) were treated with adriamycin. Adriamycin was administered according Bonadona's schedule B: the drug was injected intravenously every other day for 4—6 doses of 0.4 mg/kg and then, after a 3-day rest period, once a week.

Results

The response to adriamycin is shown in Table 1 according to KARNOFSKY's criteria. Two of 5 cases of lung cancer showed O—B. Two cases of reticulum cell

Table 1. Adriamycin in the treatment of solid tumor

Diagnosis	No. of cases	KARNOFSKY's criteria					
		O—O	O—A	O—B	O—C	1—A	1—B
Cancer of lung	5	3		2			
Cancer of larynx	1	1					
Cancer of stomach	1	1					
Cancer of rectum	1					1^{3+}	
Cancer of ovary	1	1					
Cancer of cervic	1	1					
Chondrosarcoma	1	1					
Malignant neurilemmoma	1					1^{3}	
Retroperitoneal sarcoma	1			1			
Mediastinal tumor	1						1^{2+}
Reticulosarcoma	2			2			
Total	16	8	0	5	0	2	1

sarcoma showed O—B. In a case of reticulum cell sarcoma, which was in the terminal phase and was resistant to many drugs, disappearance of lymphadenopathy was observed, but the response was transient. In another case of reticulum cell sarcoma, also in the terminal phase and resistant to all drugs, a marked decrease in number of tumor cells in bone marrow was observed.

In a case of rectal cancer with peritoneal involvement, disappearance of ascites and decrease of abdominal pain were observed for more than 3 months ($1—A^{3+}$). In a case of malignant neurilemmoma of the chest wall with multiple bone metastasis, recalcification of bone in the osteolytic humerus was observed with symptomatic improvement for 3 months ($1—A^{3}$). A case of mediastinal tumor showed the response of $1—B^{2+}$ (Fig. 1). Multiple large metastatic tumor shadows were observed (Fig. 2).

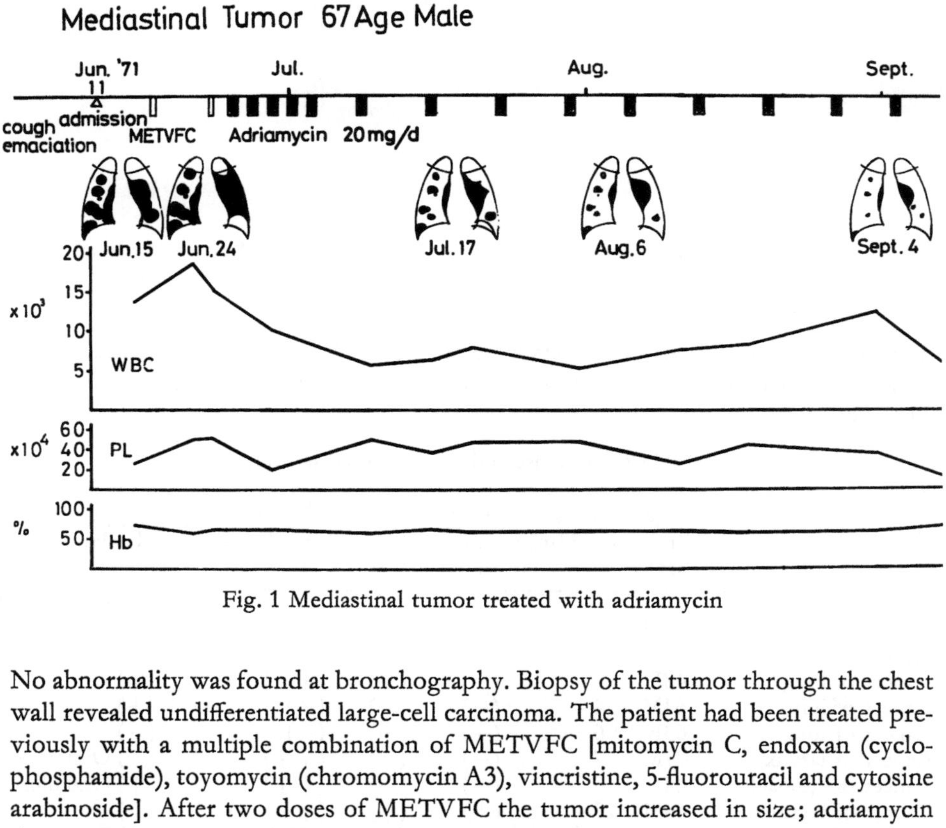

Fig. 1 Mediastinal tumor treated with adriamycin

No abnormality was found at bronchography. Biopsy of the tumor through the chest wall revealed undifferentiated large-cell carcinoma. The patient had been treated previously with a multiple combination of METVFC [mitomycin C, endoxan (cyclophosphamide), toyomycin (chromomycin A3), vincristine, 5-fluorouracil and cytosine arabinoside]. After two doses of METVFC the tumor increased in size; adriamycin therapy was then started. After 14 doses of adriamycin the tumor shadows decreased remarkably in size (Fig. 1, 3), and marked clinical improvement was observed for more than 2 months.

The response to and toxicity of adriamycin in each case are shown in Table 2. Subjective response was observed in 4 cases (25%) and objective response in 8 cases (50%). Leucocytopenia of less than 4000 mm^3 was observed in 8 cases (50%) and leucocytopenia of less than 2000 mm^3, indicated by ++, was observed in 6 cases (38%). Thrombocytopenia of less than 100000 mm^3 was observed in only 2 cases and in one case of less than 30000 mm^3, indicated by ++. Loss of hair was observed in

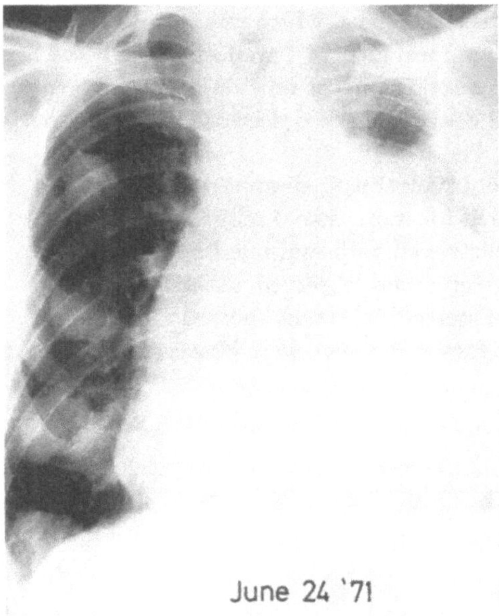

Fig. 2. Before treatment

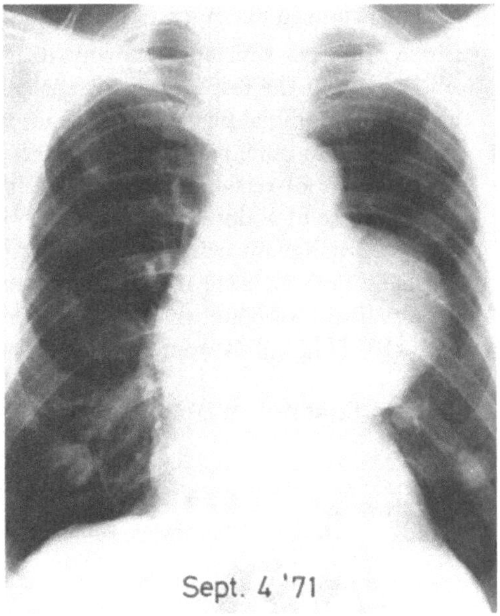

Fig. 3. After 14 doses of adriamycin (total 280 mg)

11 cases (69%), but other side effects were rare. Oral ulceration was observed in only 2 cases and diarrhea in 1 case. No nausea or vomiting and no cardiotoxicity were observed.

Table 2. Response and toxicity of adriamycin

	Age and sex	Doses (× 20 mg)	Response		Side effects		
			subjective	objective	WBC ↓	Pl ↓	Epilation
Cancer of lung	38M	5	(−)	(+)	(+)	(−)	(+)
Cancer of lung	48M	5	(−)	(−)	(−)	(−)	(−)
Cancer of lung	48F	7	(−)	(+)	(++)	(−)	(+)
Cancer of lung	64M	4	(−)	(−)	(−)	(−)	(+)
Cancer of lung	63F	6	(−)	(−)	(−)	(−)	(−)
Cancer of larynx	77M	7	(−)	(−)	(++)	(−)	(+) Oral ulcer
Cancer of stomach	44F	5	(−)	(−)	(++)	(−)	(+) Oral ulcer
Cancer of rectum	43M	15−	(+)	(+)	(++)	(−)	(+)
Cancer of ovary	55F	4	(−)	(−)	(+)	(−)	(−) Diarrhea
Cancer of cervic	54F	6	(−)	(−)	(−)	(−)	(+)
Chondroblastoma	25M	5	(−)	(−)	(−)	(+)	(+)
Malig. neurilemmoma	41M	15	(+)	(+)	(−)	(−)	(+)
Retroperit. sarcoma	43M	6	(−)	(+)	(−)	(−)	(−)
Mediastinal tumor	67M	10−	(+)	(++)	(−)	(−)	(−)
Reticulosarcoma	26F	3	(−)	(+)	(++)	(−)	(+)
Reticulosarcoma	34F	5	(+)	(++)	(++)	(++)	(+)
			4/16 (25%)	8/16 (50%)	8/16 (50%)	2/16	11/16 (69%)

Comments

These results were from an experience of a small number of cases who had previously received other treatment and were resistant to many drugs. However, an objective response was obtained with adriamycin in many cases. Leucocytopenia was observed in a relatively high percentage of cases, but they had been treated with many other drugs before adriamycin. Symptomatic side effects were relatively slight, except for loss of hair, compared with other data previously reported.

From the experimental study of combination chemotherapy reported by Hoshino, I would like to combine adriamycin with other drugs which have a synergistic effect, for example cytosine arabinoside, mitomycin C and carbazilquinone. The latest is a new chemotherapeutic agent, synthesized in Japan very recently and having a similar chemical structure to mitomycin C, it is effective against lung cancer.

Adriamycin will be one of the important drugs in combination chemotherapy of cancer.

Adriamycin, Vincristine and Prednisone in the Treatment of Malignant Tumors Resistant to Conventional Chemotherapy

H. J. SENN, W. F. JUNGI, A. C. MAYR and G. A. NAGEL

Oncology Center, Basel University Clinics, Basel/Switzerland

Introduction

Adriamycin (14-hydroxy-daunorubicine), a new cytotoxic antibiotic isolated from cultures of streptomyces peuceticus var. caesius, was introduced to clinical chemotherapy of human neoplasms in 1969 by BONADONNA and co-workers [1]. The drug seems to have significant antitumor activity in a much larger spectrum of malignant diseases than daunorubicine, especially in solid tumors [1—3]. It has mainly been used as a single agent. It was the aim of this study to explore antitumor effect and toxicity of the combination chemotherapy with adriamycin, vincristine and prednisone.

Materials and Methods

25 patients with a variety of tumors resistant to conventional protocol chemotherapy were treated at the Basel Oncology Center with the combination of adriamycin, vincristine and prednisone (Table 1). All patients had biopsy-proven, advanced, disseminated malignant disease. Classification of leukemias and staging of lymphomas as well as criteria of treatment response and toxicity conformed to the rules of the "Swiss Group for Clinical Cancer Research" and the "Acute Leukemia Group B".

Adriamycin was given at a dose of 20 mg/m^2 on days 1, 2, 3, then 11, 12, 13 and 21, 22, 23 as a rapid intravenous injection. In patients with reduced bone marrow granulocyte reserve the dose of adriamycin was reduced to 12 mg/m^2/i.v. Vincristine was administered 1 mg/m^2 i.v. on days 1, 10 and 20 and prednisone orally at a dose of 40 mg/m^2 daily for 30 days.

This 30-day induction regimen was followed by a thorough clinical and radiological evaluation of antitumor effect. In cases of clearcut remission or significant improvement, the induction was first followed by a "maintenance phase" with weekly 20 mg/m^2 of adriamycin i.v. and monthly reinductions with vincristine 1 mg/m^2/i.v. and prednisone 40 mg/m^2/p.o. for 7 days. Later in the study, this maintenance regimen was supplemented by administration of chlorambucil 4 mg/m^2/p.o. or cyclophosphamide 100 mg/m^2/p.o. daily and prolongation of the adriamycin injections to 20 mg/m^2/i.v. once a month.

Results

Table 1 shows the results of the 3-drug combination chemotherapy in 17 of 25 patients whose treatment course was evaluable at the time of this report. Remarkably good results, with partial remissions ranging from 40—60% tumor regression,

Table 1. Results of combination chemotherapy with adriamycin, vincristine and prednisone

Disease group	n	cases evaluable	CR	PR	IMP	NC	P	Too early
Acute myelocytic leukemia	5	4	0	0	0	1	3	1
Lympho-reticular sarcoma	7	6	0	5	0	0	1	1
Metastatic carcinoma	8	4	0	1	1	0	2	4
Mesenchymal tumors (sarcomas)	5	3	1	1	0	0	1	2
Total	25	17	1	7	1	1	7	8
			8/17 = 46%					

CR = complete remission, PR = partial remission (>25%), IMP = improvement, NC = no change, P = progression.

were obtained in 5/7 patients with lymphoreticular sarcomas (3 lymphosarcomas stage IV, 1 reticulum cell sarcoma stage IV and 1 lymphoepithelial tumor of the eye with multiple lung metases). No remission or even improvement was seen in 4 patients with acute myelocytic leukemia relapsing from remissions induced by VAMP or cytosine arabinoside + 6-thioguanin. Antitumor effect was also observed in 2/8 patients with metastatic carcinomas (1 PR in a patient with breast cancer resistant to hormone therapy and combination chemotherapy with cyclophosphamide, methotrexate and prednisone, 1 improvement in a patient with carcinoma of the ovary). Of special interest are the responses in 2/5 patients with soft tissue or bone sarcomas: 1 PR in a 16 year old girl with metastatic Ewing's sarcoma and a complete remission of huge pulmonary masses in a 13-year-old girl with metastatic synovialoma of the right heel. This latter response was dramatic in onset, and the patient at this time has been in complete subjective and objective remission for nearly 4 months.

The average duration of remission was 2 months, ranging from 4 weeks to 18 + weeks.

Toxicity

Table 2 summarizes the toxicity observed during the 3-drug combination chemotherapy. *Clinical toxicity* with subjective symptoms occured in all cases, with alopecia and stomatitis present in 90% and 70% respectively, most probably due to adriamycin. Gastrointestinal side effects were seen in 30% of the patients and were rather mild. Clinical infection was observed in 40%, mostly during phases of adriamycin-induced leukopenia, with 2 episodes of gram-negative sepsis most probably drug related and fatal in 1 case (Klebsiella). Neurotoxicity due to vincristine was mild and transient.

There were 2 episodes of atrial arrhythmia (1 atrial fibrillation and 1 atrial flutter) in 2 patients, occurring within several hours after a repeat injection of adriamycin and requiring interruption of therapy in one case of acute myelocytic leukemia.

Table 2. Toxicity of combination chemotherapy with adriamycin, vincristine and prednisone

Clinical toxicity	Observed in % of patients	Laboratory toxicity	Observed in % of patients
Alopecia	95	Elevation of LDH	100
Stomatitis	70	Leukopenia (marked)	70
Infections	40	Thrombopenia (mild)	30
Paresthesia (vincristine)	40	Elevation of SGPT	20
Nausea	20	Elevation of CPK	20
Cardiac (see text)	15		

Laboratory toxicity consisted mainly of leukopenia (80% of all patients), which was usually severe and the dose-limiting factor in the administration of adriamycin courses. Platelet depression was seen in 30% and was of rather mild degree. Other laboratory changes included regular elevations of LDH in serum, not conclusively due to the combination chemotherapy, and marked elevation of CPK in 4/17 patients exhibiting normal CPK levels at the start of therapy. ECG changes, except for the 2 cases of arrhythmia, included unspecific depression of the ST-segment in 2 other patients and a clearly left ventricular ischemic pattern in another patient without clinical signs of myocardial damage. Since the ECG changes did not parallel the CPK elevations except for 1 patient, a causative relationship of the cardiac changes to the chemotherapy with adriamycin, vincristine and prednisone is suspected, but not proven.

Discussion

The combination chemotherapy with adriamycin, vincristine and prednisone was found to be an effective regimen in a group of patients with various tumors resistant to or relapsing from optimal standard chemotherapy. The overall remission rate of over 40% (CR and PR) and of 70% in lymphoreticular sarcomas seems higher than expected for adriamycin alone in pretreated patients [2]. Final comparison will require a randomized study between adriamycin alone and in combination chemotherapy, which will be subject of a cooperative study of the Swiss Group for Clinical Cancer Research [4]. Vincristine and prednisone were chosen for combination chemotherapy because of their lack of additional bone marrow depression and possible increase of antitumor activity in analogy to the respective efficacy of daunorubicine, vincristine and prednisone in lymphomas and acute lymphocytic leukemias. The antitumor effect in differentiated sarcomas is of great interest and deserves further study.

Toxicity was dominated by the leukopenia, stomatitis and alopecia induced clearly by adriamycin, while neurotoxicity due to vincristine and steroid side effects were mild in the present regimen. The danger of complicating infections — although mostly transient — is a prominent feature during the induction treatment.

The main problem of the study was the short duration of remission. Neither adriamycin nor vincristine seem to be of much value in maintaining tumor regression. Whether the addition of alkylating agents and/or antimetabolites to the reinductions with adriamycin (and vincristine and prednisone) will significantly prolong remission duration, will be subject to further study [4].

References

1. Bonadonna, G., Monfardini, S., De Lena, M., Fossati-Bellani, F.: Brit. med. J. **1969 III**, 503–506.
2. Bonadonna, G., Monfardini, S., De Lena, M., Fossati-Bellani, F.: Cancer Res. **30**, 2572–2582 (1970).
3. Obrecht, P., Westerhausen, M., Simon, A.: Dtsch. med. Wschr. **96**, 369–376 (1971).
4. Swiss Group for Clinical Cancer Research, protocol 23/72.

Adriamycin was kindly supplied for this study by Deutsche Farmitalia GmbH, Freiburg i. Br./Germany.

MIX
Papier aus verantwortungsvollen Quellen
Paper from responsible sources
FSC® C105338

If you have any concerns about our products,
you can contact us on
ProductSafety@springernature.com

In case Publisher is established outside the EU,
the EU authorized representative is:
Springer Nature Customer Service Center GmbH
Europaplatz 3, 69115 Heidelberg, Germany

Printed by Libri Plureos GmbH
in Hamburg, Germany